After the Breakup

After the Breakup

U.S. Telecommunications in a More Competitive Era

ROBERT W. CRANDALL

The Brookings Institution / Washington, D.C.

1775 Massachusetts Avenue, N.W., Washington, D.C. 20036

Library of Congress Cataloging-in-Publication data

Crandall, Robert W.
After the breakup : U.S. telecommunications in a more competitive era / Robert W. Crandall.
p. cm.
Includes bibliographical references and index.
ISBN 0-8157-1606-0—ISBN 0-8157-1605-2 (pbk.)
1. Telephone—United States. 2. Telephone—United States—Deregulation. I. Title.
HE8815.C73 1991
384.6'3'0973—dc20 90-24452
CIP

9 8 7 6 5 4 3 2 1

The paper used in this publication meets the minimum requirements of the American National Standard for Information Sciences—Permanence of paper for Printed Library Materials, ANSI Z39.48–1984

THE BROOKINGS INSTITUTION

The Brooking Institution is an independent organization devoted to nonpartisan research, education, and publication in economics, government, foreign policy, and the social sciences generally. Its principal purposes are to aid in the development of sound public policies and to promote public understanding of issues of national importance.

The Institution was founded on December 8, 1927, to merge the activities of the Institute for Government Research, founded in 1916, the Institute of Economics founded in 1922, and the Robert Brookings Graduate School of Economics and Government, founded in 1924.

The Board of Trustees is responsible for the general administration of the Institution, while the immediate direction of the policies, program, and staff is vested in the President, assisted by an advisory committee of the officers and staff. The by-laws of the Institution state: "It is the function of the Trustees to make possible the conduct of scientific research, and publication, under the most favorable conditions, and to safeguard the independence of the research staff in the pursuit of their studies and in the publication of the results of such studies. It is not a part of their function to determine, control, or influence the conduct of particular investigations or the conclusions reached."

The President bears final responsibility for the decision to publish a manuscript as a Brookings book. In reaching his judgment on the competence, accuracy, and objectivity of each study, the President is advised by the director of the appropriate research program and weighs the views of a panel of expert outside readers who report to him in confidence on the quality of the work. Publication of a work signifies that it is deemed a competent treatment worthy of public consideration but does not imply endorsement of conclusions or recommendations.

The Institution maintains its position of neutrality on issues of public policy in order to safeguard the intellectual freedom of the staff. Hence interpretations or conclusions in Brookings publications should be understood to be solely those of the authors and should not be attributed to the Institution, to its trustees, officers, and other staff members, or to the organizations that support its research.

Foreword

IN 1982 THE U.S. government and AT&T entered into a consent decree that settled an eight-year-old antitrust suit. AT&T, then the world's largest corporation, would be broken up into eight separate companies in 1984, and the divested local Bell telephone companies would not be allowed to engage in the manufacture of telephone equipment, long-distance services, or information services. This move was perhaps the most far-reaching antitrust remedy since the giant *American Tobacco* and *Standard Oil* cases of 1911. But would it work? Was it wise to break up what many people called the most efficient telephone network in the world?

In this book, Robert W. Crandall analyzes the effects of the breakup of AT&T and the trend toward competition in the U.S. telecommunications sector that preceded it. He finds that the divestiture has improved productivity in the industry. Because competition has not completely replaced monopoly, however, the U.S. telecommunications sector remains highly regulated. Citing the problems that occur in regulated competition in other industries such as airlines, trucking, and railroads, Crandall points out the inherent dangers of such regulation in telecommunications as well.

Robert Crandall is a senior fellow in the Economic Studies program at Brookings. He gratefully acknowledges the suggestions and criticisms received from Gerald Brock, Charles Jackson, Almarin Phillips, Bridger Mitchell, Clifford Winston, Bruce Egan, Kenneth Flamm, Leland Johnson, Peyton Wynns, Leonard Waverman, and John Musgrave. He also thanks Menzie Chinn, Elizabeth Schneirov, Sarah Bales, Jonathan Galst, Dawn Senecal, and Christopher Owen for research assistance. Theresa Walker edited the manuscript, Pamela Plehn verified it, and Susan Woollen prepared it for typesetting. David Rossetti provided secretarial assistance. Florence Robinson prepared the index.

Brookings gratefully acknowledges partial funding for this project from the Alfred P. Sloan Foundation and the General Electric Foundation.

The views expressed here are those of the author and should not be attributed to the above individuals or to the officers, trustees, or other staff members of the Brookings Institution.

BRUCE K. MAC LAURY
President

Washington, D.C.
December 1990

Contents

Tables

Figures

After the Breakup

Chapter One

The Changing U.S. Telephone Network

THE EXPLOSION OF electronics technology over the past fifty years shows little sign of abating. This technological revolution, affecting communications, has expanded the scope of telephone service beyond anyone's wildest dreams. The typical U.S. caller may now dial almost anywhere in the world directly, store and forward a message, or even transmit a facsimile reproduction of any document in less than a minute, often for less than the real cost of a 500-mile telephone call twenty-five years ago.

Dramatic changes in government regulation, pricing, and competition have accompanied these technological improvements in the telecommunications industry, sometimes confusing and alarming consumers. As a result of the 1982 antitrust decree that divided the American Telephone and Telegraph Company (AT&T) into eight separate firms and spun off two other telephone companies in which AT&T had a minority interest, residential customers can no longer buy local service, long-distance service, and telephone equipment from the same vendor.[1] The telephone equipment and services sector has changed from a tranquil, regulated monopoly into a set of increasingly competitive markets in which domestic and foreign suppliers compete for the patronage of household and business users.

Consumers are also facing more detailed and complex monthly telephone bills. The charge for residential local service now includes a federally mandated subscriber-line charge of $3.50 a month besides the usual flat monthly rate. Moreover, measured local service is more common than in earlier decades, and long-distance

1. *United States* v. *American Telephone and Telegraph Co.*, 552 F. Supp. 131 (D.D.C. 1982), *aff'd. sub nom., Maryland* v. *United States*, 460 U.S. 1001, 103 S. Ct. 1240, 75 L. Ed. 2d 472 (1983).

charges are usually billed separately for short-haul intrastate calls and for longer interstate and intrastate calls. Some of these long-distance rates have fallen dramatically while others have declined very little, and local rates have risen sharply since 1981.

Federal and state regulators, as well as a U.S. district court that administers the 1982 antitrust decree, still constrain much of the new telecommunications rivalry. Indeed, many observers believe that the confusion created by overlapping regulatory jurisdictions and the puzzling array of competitive services and equipment more than offsets the benefits of new services and potentially lower prices.[2] The array of choices available and the mass of detail in telephone bills may confuse customers, and they may even fear that the introduction of competition has destroyed the integrity of the telephone network and raised the cost of gaining access to it. Moreover, economists rightly fear that competition and rate-of-return regulation are an unhealthy combination, threatening to induce industry cartelization and the protection of inefficient competitors.[3]

The Evolution of Telephone Service

In the late nineteenth century, the telephone network was composed only of pairs of copper wires strung between households or businesses and telephone offices that allowed the transmission of electrical signals that could be encoded and decoded into voice messages. Most telephone subscribers are still connected to the network by twisted pairs of copper wires.

Voice messages are routed from one customer to another by "switches," devices that allow a caller's circuit to be connected to the intended receiver's circuit. At first these switches were little more than mechanically automated versions of a switchboard, but shortly after World War I switches evolved into complex, automated, electromechanical devices. Today, switching is handled by large

2. There are many recent books critical of the changes in U.S. telecommunications industry structure. See, for example, Alan Stone, *Wrong Number: The Breakup of AT&T* (Basic Books, 1989); or Robert Britt Horwitz, *The Irony of Regulatory Reform: The Deregulation of American Telecommunications* (Oxford University Press, 1989).

3. John Haring, "The FCC, the OCCs and the Exploitation of Affection," OPP Working Paper 17 (Washington: Federal Communications Commission, June 1985).

computers—referred to as "stored program control, time-division switches."[4] Switches of various designs route local calls within an exchange, to trunks between exchanges, and to interexchange trunks for long-distance calls. In addition, the switches perform various control and record-keeping functions.

Obviously, not all telephone subscribers can be connected to the same switch. The cost of connecting any subscriber to a switch increases with distance from the switch. A trade-off always exists between increasing these connection costs by reaching out for more distant subscribers and the economies available from switching them through a single switch. A single metropolitan area may have dozens of local switches organized around local exchanges, each one identified by the first three numbers in a local telephone number.

The telephone operating companies, or local-exchange carriers (LECs), operate many exchanges in various cities, towns, and rural areas. These companies offer a variety of basic services: access service—the access to the telephone network; local-exchange service—the connection of calls in a given local-calling area; and interexchange service—the connection of calls to customers served by customers outside the caller's local-calling area.

In practice, little difference exists between local-exchange and interexchange service. A local call is often routed between two or more switching machines in different exchanges in the same metropolitan area. A long-distance call is simply transmitted a little farther between exchanges. In the United States, many local exchange carriers have traditionally provided some long-distance service. Even today, after the breakup of AT&T, the divested Bell operating companies provide considerable intrastate long-distance service. AT&T and several newer long-distance carriers, such as MCI and Sprint, offer interstate long-distance service as well as intrastate service in some states.

Telephone messages may now be transmitted over many different facilities. Local calls from residential or small commercial and industrial facilities are still largely transmitted to the local switch over copper wires. However, as the number of voice circuits or the

4. For a description of the elements of the telephone network, see R.F. Rey, *Engineering and Operations in the Bell System*, 2d ed. (Murray Hill, N.J.: AT&T Bell Laboratories, 1983).

amount of data transmitted over any local route increases, the signals may be transmitted by microwave facilities or even by optical fibers. Long-distance calls may be transmitted by microwave, satellite, or optical fibers. In recent years, optical fibers have become dominant in long-distance services because of their enormous capacity, lower and declining real costs, and signal clarity.

The development of microwave during World War II allowed new competitors to enter the long-distance telephone business because the economies of scale in microwave transmission were small compared with the size of many interexchange markets. Satellite transmission is now used largely for data or for broadcast signals and for voice traffic in remote areas. Fiber optics, the newest technology, utilizes laser transmission over very fine silicon fibers and seems to offer rather substantial economies of scale. As a result, the shift to fiber optics by AT&T and the new long-distance competitors may ironically lead to renewed market concentration in long-distance services.

Originally, voice messages were transmitted in analog form. However, with the development of large-scale, solid-state electronics technology, telephone service began to evolve into a voice and data service in the 1960s. As a result, digital conversion of all signals became more efficient. The older analog telephone systems with crossbar or step-by-step switches were subject to potentially high error rates in transmitting bits of data and are being replaced by digital circuits.

The high costs and unreliable service from older analog telephone systems led large data users to seek alternative modes for transmitting their data. Private microwave was first licensed in 1959 and specialized carriers were authorized by the Federal Communications Commission (FCC) beginning in 1969.[5] These new specialized carriers had to use local telephone company circuits to gain access to most of their customers and to terminate their calls because they owned no local circuits. In a few cases, very large customers could utilize direct microwave links for access, but these circuits that bypassed the local telephone companies were relatively rare.

The distinction between voice and data transmission is now beginning to blur. Most telephone companies are converting entirely

5. See chap. 2.

to digital technology. Customers may want to transmit voice messages, send data from point-of-sale terminals, send and receive facsimile transmissions, reroute voice and data signals to other receiving points when the intended receiver is busy, and conduct other business over telephone circuits. As a result, telephone companies are now busy developing Integrated Services Digital Networks that offer customers a wide array of services through a single integrated network using digital technology.

Telecommunications is rapidly moving beyond the mere transmission of data or voice messages. Given current technologies, telephone carriers can provide services that would have been impossible even twenty-five years ago. These services include the storage and tranformation of data for households or businesses. In the current rubric of telephone regulation, these services have become known as information services.

Among the most simple of the new information services is code and protocol conversion necessary to allow dissimilar computers or terminals to communicate with each other. Other services involve the use of computers for storage or data processing. Still others simply provide for the remote storage and accessing of messages. And perhaps the most important are transactions services that allow banks to provide remote transaction terminals, merchants to process credit card purchases and verify a customer's creditworthiness, or others to provide shop-by-phone services.

The most well-known of the current information services are the videotex services that allow consumers to engage in transactions or information searches through personal computers or video terminals connected to telephone lines. The French postal, telegraph, and telephone authority (PTT) has been quite successful in developing its Teletel system by subsidizing (Minitel) terminal installations and offering telephone directory services by videotex as a substitute for printed directories.[6]

The AT&T antitrust consent decree forbids the divested Bell operating companies to offer information services. This prohibition has recently been relaxed slightly to allow the Bell operating companies to provide the gateway circuits for videotex systems and

6. For a more complete discussion of information services, see National Telecommunications and Information Administration, *Information Services Report*, NTIA Report 88–235 (Department of Commerce, August 1988).

voice-storage services, but the operating companies still may not offer content-based information services.[7]

Telephone Equipment

The early telephone monopoly was based on patents for the basic telephone transmitting and receiving equipment.[8] Although many of these patents expired in the late nineteenth century, U.S. telephone equipment manufacture remained mostly in the hands of telephone operating companies (or their holding companies) until the 1980s.

Telephone equipment may be divided into three broad categories: transmission equipment, switching equipment, and terminal equipment. Transmission equipment was at first confined to the transmission of electrical pulses over wires. Today, however, signals are transmitted over the electromagnetic spectrum by microwave equipment and by satellites, through large coaxial cables, and through optical fibers. In addition, a variety of radio-based equipment transmits and receives mobile telephone calls, often through complex cellular networks.

Switching equipment may be located in a local company exchange office, a long-distance carrier's facilities, or even in a subscriber's facilities. The switches located in telephone company exchange offices not only switch local calls, but they also route long-distance calls to long-distance carriers or to other exchanges in the telephone company's franchise area. They also record call completion, the duration of usage, and the customer's billing information. Recent developments in electronics technology have revolutionized telephone switching, allowing central office switches to perform several functions, such as message storage or call forwarding, and to offer equivalent connections to numerous long-distance companies.

Not all switches are located in telephone-company offices. Customers may also own smaller switches to transfer calls among offices or apartments in a large complex or even among buildings. The

7. See *MFJ Modification Order, United States* v. *Western Electric*, 714 F. Supp. 1 (D.D.C. 1988).

8. See Gerald W. Brock, *The Telecommunications Industry: The Dynamics of Market Structure* (Harvard University Press, 1981); and chap. 2, for a more complete discussion.

larger switches are called private branch exchanges (PBXs) and are similar to a telephone company switch.[9] Indeed, customer switching is an alternative to telephone-company switching for many large customers and is a potent force in constraining rising telephone rates. For example, PBXs may be connected directly to a long-distance carrier or to other PBXs, thereby bypassing a local carrier's circuits altogether.

Terminal equipment is the vast array of equipment used by customers to originate or terminate calls. At first, terminal equipment was simply a handset. However, even the simple handset has undergone a startling evolution, first to a rotary dial, then to push-button or touch-tone signaling, and more recently to a radio-based cordless mode or even a truly portable transceiver. Key telephones are telephone sets with keys that allow the access to multiple lines and even multiple services, such as call forwarding, holding, conferencing, and other services. More complex key systems permit some switching outside the central office switch but are less complex than full PBXs.

The electronics revolution has created the demand for new voice, data, and record services and the attendant growth of diverse types of terminal equipment. Modems provide the conversion of computer data into a format suitable for transmission over the telephone network. Facsimile machines allow written documents to be transmitted and received over telephone lines and are now replacing older Telex equipment. A variety of other equipment has developed to link computer networks to the public telephone network.

The U.S. Telephone Industry

The total telephone and telegraph services industry is immense. AT&T and each of the regional Bell companies rank among the country's largest one hundred firms. In 1988 total industry revenues were more than $100 billion, of which long distance accounted for 55 percent, local service for 36 percent, and telegraph, Yellow Pages,

9. The modern, digital form of these switches are often referred to as private automatic branch exchanges.

Table 1-1. U.S. Telephone and Telegraph Industry Revenues, 1988
Billions of current dollars

Telephone service	*Revenues*
Total	111.9
Local telephone[a]	40.4
Long-distance telephone[a]	61.6
Interstate	38.2
Intrastate	23.4
Miscellaneous telephone[a]	9.0
Telegraph	0.9

Sources: Author's calculations based on United States Telephone Association (USTA), *Statistics of the Local Exchange Carriers 1989* (Washington, 1989); and Federal Communications Commission (FCC), *Statistics of Communications Common Carriers* (Washington, 1988–89), telegraph industry revenues from table 1-4.
a. Estimate based on FCC detailed data and USTA data for total revenues.

and other services the remaining 9 percent (table 1-1). Revenues have shifted steadily from local to long-distance over the past few decades. For instance, in 1960, long-distance revenues accounted for only 36 percent of revenues while local revenues were 56 percent of total revenues. This shift has occurred despite the relative rise in local rates in recent years because of the repricing of access service induced by federal regulation.

Until the 1960s, the U.S. telephone industry was composed of three groups of firms: AT&T and its associated companies, the independent companies, and rural telephone cooperatives. AT&T owned the Bell operating companies that served roughly 80 percent of the nation's telephone subscribers. The independent companies, including those affiliated with Continental Telephone, United Telephone, and GTE, accounted for most of the rest. Many small, rural telephone companies, including numerous cooperatives, subsidized by Rural Electrification Administration and Rural Telephone Board low-interest loans, offered service to about 4 percent of the nation's subscribers.

In 1974 the Justice Department filed an antitrust suit against AT&T, alleging that AT&T had monopolized telephone equipment manufacture and long-distance service through its control of local exchanges. This suit was settled by a consent decree entered in the U.S. district court in 1982, requiring AT&T to spin off its telephone operating companies in 1984.[10] AT&T was allowed to keep its

10. Technically, this decree was entered as a modification of the final judgment (MFJ) that settled an earlier antitrust case in 1956.

Long Lines Division, its Western Electric telephone-equipment manufacturing division, and most of Bell Laboratories.

The breakup of AT&T led to the creation of seven regional Bell holding companies, which operate the divested Bell operating companies, and to the separation of two other large operating companies in which AT&T had a large interest. These companies continue to provide local telephone service to 80 percent of the nation's telephone access lines (table 1-2). The remaining 20 percent are served by the independent companies, including the rural coops. Under the terms of the AT&T consent decree, the divested regional holding companies may not offer long-distance service outside their local access and transport areas (LATAs), must refrain from engaging in the manufacture of equipment, and may not offer content-based information services.

Until the 1970s, AT&T provided virtually all interstate long-distance service. AT&T's Long Lines and the Bell operating companies offered switched-voice and data long-distance services as well as dedicated private-line services to large users and even provided most of the interconnecting services for radio and television broadcasting. In 1969 the Federal Communications Commission began to admit new competitors into interstate long-distance service on a limited basis. By the late 1970s, entry into all interstate interexchange services was permitted, and new competitors, such as MCI, Sprint, and numerous regional carriers began to proliferate.[11] In addition, the FCC began to admit resellers of AT&T bulk services—or arbitragers—who thrived on the low bulk-rate tariffs offered by AT&T in response to competitive pressures.

The interstate (and international) long-distance market is much more competitive today than before the AT&T divestiture. By 1988 AT&T had slightly more than 75 percent of the market; MCI and U.S. Sprint had about 18 percent; and several smaller carriers shared the remaining 7 percent (table 1-3).

The divorce of the Bell operating companies from AT&T severed the most important connection between the exchange carriers and equipment manufacturing. Under the 1982 decree, the divested Bell operating companies are forbidden to manufacture equipment. As

11. The evolution of the interstate market from AT&T monopoly to competition is treated in greater detail in chap. 2.

Table 1-2. The Largest U.S. Telephone Operating Companies, 1988

Company	*Access lines (thousands)*	*1986 operating revenues (millions of dollars)*
Regional Bell holding companies		
Bell Atlantic Corporation	16,541	9,730.3
Bell of Pennsylvania	4,979	2,732.8
Chesapeake and Potomac Telephone Companies	6,529	4,111.9
Diamond State Telephone	389	224.2
New Jersey Bell Telephone Company	4,643	2,857.9
Bell South Corporation	16,407	11,806.1
South Central Bell	6,977	4,784.6
Southern Bell	9,430	6,748.6
American Information Technologies Corporation	15,469	9,068.3
Illinois Bell	5,084	2,712.3
Indiana Bell	1,563	976.1
Michigan Bell	4,046	2,458.9
Ohio Bell	3,138	1,992.9
Wisconsin Bell	1,638	978.7
Nynex Corporation	14,851	11,111.2
New England Telephone and Telegraph	5,432	3,612.0
New York Telephone Company	9,668	7,481.1
Pacific Telesis Group	13,093	8,906.4
Nevada Bell	205	157.3
Pacific Bell	13,813	8,062.2
Southwestern Bell Corporation	11,340	7,200.3
U.S. West	11,878	8,078.9
Mountain Bell	5,901	3,772.5
Northwestern Bell	3,586	2,304.5
Pacific Northwest Bell	2,765	2,012.3
Independent holding companies		
GTE Corporation (U.S. only)	12,180	9,617.0
United Telecommunications, Inc.	3,685	2,509.7
Contel Service Corporation	2,470	1,989.7
Southern New England Telephone Company	1,839	1,368.1
Centel Corporation	1,503	858.4
Alltell	1,084	733.6

Source: USTA, *Statistics of the Local Exchange Carriers, 1989*, pp. 6, 15–37.

a result, for the first time in the history of the U.S. telephone industry, most telephone equipment is produced by firms that are independent from local telephone exchange operations. Domestic switching and transmission equipment is now manufactured by many companies for the U.S. market although large central office switches are still mostly the province of old-line telephone equipment manufacturers, such as AT&T, Northern Telecom (Canada), GTE, Ericsson (Sweden), Siemens (Germany), and NEC (Japan).

Table 1-3. Long-Distance Carriers, 1988
Millions of current dollars

Company	*Revenues*[a]
AT&T	35,407
MCI	4,886
U.S. Sprint	3,405
Telecom U.S.A.	524
Allnet	394
USTS	282
Cable & Wireless	218
Advanced Telecom Corp.	178
Others[b]	1,654
Total	46,949

Source: FCC, *Statistics*, table 1-3.
a. These revenues are largely interstate long-distance revenues. Intrastate long distance is still dominated by the local-exchange carriers.
b. Others includes Alascom.

Terminal equipment manufacture has become a very competitive industry since telephone-company customers were given the right to attach their own equipment to their telephone lines in the late 1970s. Computer companies, traditional large telephone equipment companies, and many smaller producers throughout the world manufacture a vast array of terminal equipment for the U.S. market.

Private Carriage

Before 1959 no one except a regulated telephone or telegraph company could offer standard telecommunications services, and no one could even build his own network or own his own terminal equipment. In 1959 the FCC opened the door a crack by allowing private microwave services.[12] A decade later, these microwave services could be offered for specialized uses in common carriage. In the 1970s, terminal equipment was liberalized, and subscribers could begin owning telephone sets or private branch exchanges.

By the 1980s, large private users could build private networks from their own microwave circuits, terminal equipment, or private satellite dishes if common carriers could not offer the services they required at rates equal to their own potential costs, or they could

12. To obtain a license for private microwave, a user had to demonstrate the absence of common-carrier alternatives.

augment common carrier services with some of their own equipment. Airline reservation systems, large banking networks, and private networks connecting several large office buildings or even a multitude of retail outlets became common. A recent market research survey estimated that $7.7 billion was spent on private networks in 1988.[13]

As these large private networks have proliferated, they have posed regulatory problems because they may choose to interconnect with the public network at many locations depending on local rates, may bypass the local exchange altogether, and may even begin to offer common carrier services in competition with regulated carriers.[14] Given that a large share of common carrier costs are sunk, some regulators fear that such bypass may drive up the rates for those remaining on the network, thereby encouraging further attempts to bypass the network.

Regulation

The United States is among the few countries in the world with a private telephone system. It is unique in the degree of fragmentation and competition permitted in its telephone network.[15] In most countries, a government-owned PTT—postal, telephone, and telegraph authority—offers the entire array of local access, local exchange, and long-distance services. These PTTs usually buy their equipment from a narrow range of national equipment suppliers who do not have to suffer the discipline of truly competitive procurement practices. Usually, the PTT prohibits the interconnection of non-company-owned terminal equipment to a subscriber's line.

The U.S. telephone system is regulated by both state public utility (service) commissions and the Federal Communications Commission. Although the formal jurisdictions of these regulatory

13. Dataquest estimates published in *Communications Week*, May 15, 1989, p. 1.

14. Some large private network owners are Sears, K Mart, and Japan Air Lines. See *Communications Week*, various issues, 1988, 1989.

15. Chuck Jackson has suggested that New Zealand may have exceeded the United States in liberalizing telecommunications, having fully deregulated the sector in April 1989.

bodies are well delineated geographically, many problems and even more numerous conflicts are caused by the intangible nature of telephone service and the common costs of services crossing these boundaries. Since 1982 a third regulator—the U.S. district court for the District of Columbia has been added to this array of often contentious regulators because of the AT&T antitrust decree.

State regulators have jurisdiction over local service and intrastate long-distance markets, but the equipment used by telephone carriers for these intrastate markets is also used for interstate service. For instance, the line connecting the subscriber to the local switch is used for all of that customer's services, intrastate and interstate. Because the costs of transmission and switching have fallen more rapidly than the cost of installing and maintaining local circuits, long-distance rates could have fallen much more rapidly than local rates. However, state and federal regulators shifted a larger and larger share of the essentially fixed (or non-traffic-sensitive) costs of local circuits to the interstate long-distance service for ratemaking purposes. The result was to slow the decline in long-distance rates and keep them above costs until new entry appeared in the late 1960s and 1970s.

In the 1960s, the federal regulators, prodded by the courts, began to admit competition into service and equipment markets. Each of these competitive thrusts was generally opposed by most state regulators who feared that competition would drive rates to costs, thereby raising the customer's local monthly rate while reducing long-distance rates. In the end, the federal regulators won a modest victory, imposing fixed, monthly subscriber-line charges to defray part of the non-traffic-sensitive costs of the network.

Besides the implicit subsidy of local service caused by the assignment of a disproportionate share of non-traffic-sensitive costs to interstate calls, there has been a tradition of subsidizing rural subscribers at the expense of urban subscribers. Part of this subsidy has been direct—in the form of loans with low interest rates from federal agencies. However, much of the subsidy has derived from the formula for federal-state cost allocations and from state regulatory decisions concerning the rate structures of local-exchange companies. As a result, urban subscribers, particularly urban business subscribers, have often been forced to subsidize rural telephone service.

The distortions caused by rate regulation are far too many and too complicated to be reviewed in depth here.[16] However, earlier experience with regulated competition in transportation cautions that attempts to admit competition while maintaining regulation result in regulators cartelizing an industry to protect economically inefficient but politically powerful competitors. Rate-of-return regulation may also induce inefficient input choices, slow technological progress, and induce attempts to cross subsidize unregulated activities from a regulated market.

When the 1982 AT&T consent decree was entered, there was an expectation that equipment manufacture and long-distance service would be fully deregulated. Deregulation of equipment is now essentially complete. Its manufacture was never explicitly regulated, but the connection of terminal equipment to the telephone network was regulated before the late 1970s.[17] Interstate long-distance service, however, remains a regulated service although only AT&T's rates and services are explicitly regulated. Intrastate long-distance services are also still regulated to varying degrees by the various state commissions.

Local service remains tightly regulated by state commissions, but indications are that some state regulators may begin to consider liberalization of entry conditions and rate regulation. Even this slight opening is surprising in light of the state regulatory climate only a decade ago.

The Current Environment

The telephone industry has evolved into a much more competitive industry, constrained by federal, state, and judicial regulators. Of these three regulators, the FCC has pressed most actively for substituting competition for regulation whenever possible and for using more rational approaches to regulatory ratemaking whenever competition is economically or politically impossible. These policies have led to a significant repricing of telephone service—raising local access rates closer to costs and reducing long-distance rates markedly. Many state regulators as well as state and federal

16. See chap. 2 for a more thorough description of some of these distortions.

17. Equipment must be certified by the Federal Communications Commission as not harmful to the network, but otherwise the design and terms of sale or rental of this equipment are unregulated.

legislators have attempted to slow the pace of deregulation and cost-based pricing because of the apparently mass popular appeal of low monthly access rates. They have had limited success at best, owing to the perseverance of the FCC.

In the new competitive world, the divested Bell operating companies are the most constrained of the players, yet they account for about 60 percent of all telephone common carrier assets. The court administering the AT&T antitrust decree has recently shown some leniency in relaxing some information service restrictions on these operating companies, but they are unlikely to be allowed into long-distance service or equipment manufacturing any time soon. Moreover, the court has maintained indefinite jurisdiction over the decree, thereby constraining the other regulators and creating venues in which contesting parties may pursue their own economic interests.

Chapter Two

From Regulation to Competition to Divestiture

IN 1974 the Justice Department brought an antitrust suit against AT&T, alleging that the operator of the world's largest integrated telephone network was monopolizing the market for telephone equipment and long-distance service. The government charged that as long as AT&T controlled the local circuits that provided the only access to most consumers, competition could not flourish in long-distance service, data services, private branch exchanges, key telephone systems, large telephone switching machines, or other telephone equipment and services.

In 1982, to settle the suit, AT&T agreed to spin off its telephone operating companies into seven independent regional holding companies. These seven new entities, in turn, were forbidden to offer long-distance information services and were barred from equipment manufacture. On January 1, 1984, the Bell telephone system expired. Many critics abhorred the decision to break up AT&T. They saw telephony as a natural monopoly, enabling the company to offer the public great economies of scale and scope.[1]

Others saw divestiture as only a necessary step along the road to a more competitive market, begun in the 1960s and 1970s when the Federal Communications Commission (FCC) first allowed competition to enter the market for telephone services and equipment.[2]

How the government forced the breakup of AT&T begins with

1. See Almarin Phillips, "The Impossibility of Competition in Telecommunications: Public Policy Gone Awry," in Michael A. Crew, ed., *Regulatory Reform and Public Utilities* (Lexington Books, 1982), pp. 7–33.

2. Roger G. Noll and Bruce M. Owen, "The Anticompetitive Uses of Regulation: United States v. AT&T," in John E. Kwoka, Jr., and Lawrence J. White, eds., *The Antitrust Revolution* (Scott Foresman and Company, 1988), pp. 290–337.

the story of the industry's evolution from monopoly to competition, to regulated monopoly and, most recently, to regulated competition. No one planned this course of events. Indeed, most of the recent policy changes did not even reflect a clear, societal consensus.

The Early Days of AT&T

AT&T's early power derived from its patents on the basic technology of transmitting voice signals over copper wires, which allowed AT&T's forerunner company to establish local telephone exchange monopolies through a series of license agreements.[3] When these patents expired in the 1890s, a rush of new entry into local telephone service took place, but the new entrants did not have access to a service that would interconnect their exchanges with distant exchanges in other cities or states. AT&T had pioneered the development of long-distance technology, and it used its patents in this field to control the interexchange (long-distance) business.

In the early part of this century, AT&T used this long-distance monopoly as leverage to acquire competitive local exchange companies. Without access to a long-distance service, the independent local companies found it difficult to compete and to resist AT&T's offers to acquire them. This merger activity attracted the attention of the Justice Department, and to head off a potential antitrust suit, AT&T agreed in 1913 (the Kingsbury Commitment) to cease acquisitions of direct competitors, to interconnect independent local companies with its long-distance network, and to dispose of its Western Union stock.[4]

After Congress liberalized the antimerger law for telephone mergers in 1921, AT&T increased its share of the local exchange business to 80 percent of the country's telephones. Thus, by the late 1930s, AT&T controlled almost the entire long-distance market, 80 percent of the local exchange market, and a very large share of the market for switching and terminal equipment through its Western

3. For a more detailed analysis, see Gerald W. Brock, *The Telecommunications Industry: The Dynamics of Market Structure* (Harvard University Press, 1981), or Peter Temin, with Louis Galambos, *The Fall of the Bell System: A Study in Prices and Politics* (Cambridge University Press, 1987).

4. The Kingsbury Commitment was named after an AT&T official who negotiated the informal understanding with the Department of Justice.

Electric manufacturing subsidiary. In addition, telephone revenues from the Bell operating companies' license agreements supported the Bell Laboratories, a research organization without peer in the U.S. private sector.

Regulation

In 1910 the Interstate Commerce Commission began informal regulation of AT&T's interstate business. The states began to establish formal regulatory commissions to control intrastate telephone rates (as well as other utility rates) in the years after World War I. Finally, Congress transferred interstate telephone regulation to a new independent commission, the FCC, created by the 1934 Communications Act.

Even under the new FCC, federal regulation of AT&T could hardly be described as stringent before World War II. Individual rates were not regulated; AT&T was merely required to satisfy a broad "just and reasonable" standard for its entire interstate operations. When the FCC staff, in response to the 1934 act's requirements, began to worry that AT&T might be purchasing capital equipment from its own unregulated, equipment-supplying subsidiary, Western Electric, at supracompetitive prices, its 1938 report was suppressed because of pressure from AT&T through other government departments.[5]

The interstate telephone market before World War II was much simpler than in the years immediately following the implementation of microwave as the dominant medium for interstate transmission. Only when categories for services and rates proliferated and competitors emerged in the 1960s did AT&T's interstate rates become a matter of political and, therefore, regulatory concern.

Antitrust - I

In 1938 AT&T repulsed the first attack on its relationship with Western Electric by securing the burial of the draft FCC staff report. After World War II, however, the political impetus for antimonopoly cases induced the Justice Department to file a Section 2 Sherman Act suit against AT&T. This suit, filed in 1949, focused on AT&T's

5. Brock, *Telecommunications.*

alleged monopolization of telephone equipment through its exclusive purchases from Western Electric.[6] After a change in administrations, several Cabinet departments brought pressure on the Justice Department to settle this suit with little apparent harm to AT&T. The consent decree entered in 1956 to settle the case permitted AT&T to keep Western Electric but also prohibited AT&T from entering any markets other than regulated telecommunications.[7]

Competition in Services

No sooner had the Justice Department settled its suit against AT&T than another arm of the government, the FCC, began a rulemaking that would open the floodgates of competition and eventually destroy the integrated AT&T network. In 1956 the FCC began to investigate the possibility of allocating electromagnetic spectrum to private microwave users. In 1959 this proceeding was concluded with a decision to allow large private users to build their own microwave systems.[8] Though this decision may have seemed innocuous at first, it was the beginning of the end of AT&T's monopoly over long-distance services.

In 1961 AT&T began to respond to the competition from private microwave by filing a series of private-line tariffs with the FCC, known as Telpak, granting large users sizable discounts.[9] These tariffs were controversial because of their boldness—they reflected as much as an 85 percent discount from the single-line rates in force. More important, they set in motion proceedings that demonstrated the impossibility of regulating rates with any precision. For nearly twenty years, the FCC struggled with a succession of AT&T private-line rate filings without being able to satisfy itself that any of the successive tariffs was justified. The commission could not devise a satisfactory method for determining the costs of

6. *United States* v. *Western Electric Co., Inc. and American Telephone and Telegraph Company*, Civil Action 17-49 (D.N.J., 1949).

7. The modification of this decree formally constitutes the 1982 modified final judgment or MFJ.

8. *Allocation of Frequencies in the Bands above* 890*MC*, 27 FCC 359 (1959).

9. A private line is a dedicated line for communications between two fixed points—usually offices or other facilities of a large customer.

individual services,[10] and AT&T either would not or could not provide the data that would allow the commission to conclude that any single tariff was justified.[11]

The FCC should have taken these problems of regulating AT&T tariffs more seriously when it began to allow new firms the right to offer a competitive, private-line service to third parties. The 1959 Above-890 decision had legalized private microwave for a firm's own use. Now others were asking for the right to build microwave networks to sell services to the smaller firms that could not justify building private systems for their own limited use. In 1969, after six years of pleading with the FCC, MCI was authorized to begin building a common-carrier network for private-line services.[12] The FCC allowed this entry on the dubious premise that the new service was unique and would not compete importantly with AT&T's bread and butter: switched-voice, long-distance services.

In 1971 the FCC extended the entry authority generally to "specialized" common carriers who were expected to concentrate primarily on offering data service to smaller companies.[13] The FCC decision showed no recognition that the commission would have to regulate AT&T's individual rates (since AT&T would surely respond to this competition, and the new competitors would just as surely complain about this response). More important, the FCC failed to recognize that it would have to begin arbitrating disputes over how the new carriers would gain access to their customers. The

10. The Federal Communications Commission (FCC) and AT&T clashed over the appropriate costing methodology for pricing services delivered over common facilities. The FCC eventually settled on a "fully distributed" (FDC) historical cost standard over AT&T's objections. AT&T wanted to use long-run incremental costs (LRIC) as the basis for ratemaking—a theoretically superior methodology—but the commission distrusted this approach because it would allow AT&T too much discretion in measuring costs and because universal LRIC-based rates would not cover the AT&T revenue requirement. For a discussion of FDC pricing, see Ronald R. Braeutigam, "An Analysis of Fully Distributed Cost Pricing in Regulated Industries," *Bell Journal of Economics*, vol.11 (Spring 1980), pp. 182–96.

11. For a retrospective view of this problem from the perspective of a regulator who participated in the process, see Walter G. Bolter, "The FCC's Selection of a 'Proper' Costing Standard after Fifteen Years—What Can We Learn from Docket 18128?" in *Assessing New Pricing Concepts in Public Utilities, Proceedings of the Institute of Public Utilities Ninth Annual Conference* (Michigan State University, Graduate School of Business, 1978), pp. 333–72.

12. *Microwave Communications, Inc.*, 18 FCC 2d 953 (1969). The FCC apparently thought that its authorization of MCI's entry was limited to private-line service, but it never explicitly barred MCI from switched services. See pp. 21–22.

13. *Specialized Common Carrier Service*, 29 FCC 2d 870 (1971).

specialized carriers owned no local access lines; the local telephone operating companies controlled these lines. Coincidentally, their competitor, AT&T, owned the operating companies that controlled the access to 80 percent of the nation's telephones. Was AT&T likely to grant its new competitors easy access to these customers?

AT&T's response to this new competition took several forms. First, AT&T began to file tariffs that offered discounts to large private-line customers, and it attempted to lower rates on the more dense (and therefore, lower-cost) routes. The new competitors, who hoped to thrive on AT&T's inefficient earlier pricing, opposed these filings, and these rates were generally disapproved by the FCC, which could not determine AT&T's individual service costs on any route.[14] In addition, AT&T promoted WATS (wide-area telephone service), a discounted bulk-offering of long-distance service, more intensively as a means of offering lower long-distance rates to larger customers.

Second, AT&T resisted requests by MCI in particular for services required to connect MCI to AT&T's local customers. This resistance became a serious issue in the 1974 antitrust suit.[15]

Third, AT&T tried a legislative end run around the FCC, seeking passage of a bill that would have barred competition in interexchange markets as well as terminal equipment markets.

In retrospect, none of the AT&T strategies was very successful, although each may have slowed the development of competition before the 1980s. But even AT&T was not prepared for its largest defeat—the entry of the new carriers into the market for ordinary (switched) long-distance services. Until 1974, the FCC thought that its decision on specialized carriers relegated the new carriers to the private-line market—the market for fixed, point-to-point service used by larger businesses—in which these new carriers could provide a useful complement or spur to AT&T by developing digital data transmission services on dedicated lines for business customers. But more than 90 percent of the revenues from interexchange service

14. The FCC could only suspend the tariffs for a few months while it launched an investigation into the reasonableness of the rates. Thereafter, the tariffs would become effective. The FCC would usually take one to two years to reach a decision on the reasonableness of the tariffs, but by this time AT&T would probably wish to file new tariffs anyway. Thus, the FCC had only limited powers to control AT&T rates through this process.

15. See Judge Greene's opinion on AT&T's motion for summary judgment in the 1974 suit. (*United States* v. *American Telephone and Telegraph Co.*, 524 F. Supp. 1336 (1981).

was in the garden-variety, switched-voice message service, a market thought to be a natural monopoly.[16]

In 1974 MCI began to offer a form of switched-message service, using its private lines and the circuits that it leased from the operating companies. Although it had initially leased these circuits to terminate fixed, point-to-point service, MCI found that it could use these local connections to dial up any other local subscriber in each of the exchange areas in which it had leased the circuits. Therefore, it decided to begin to promote an Execunet service that allowed private-line customers to obtain switched long-distance service.

After AT&T complained that MCI's new service was not allowed by its FCC license, the commission attempted to enjoin MCI from offering it. However, the commission lost every court ruling in trying to block the Execunet service because the courts found that the FCC had failed to establish a record that justified limiting MCI to private-line service.[17] Had the commission found explicitly that AT&T's switched services constituted a natural monopoly, thereby requiring that entrants be barred from them, the courts would probably have deferred to the FCC's expertise in the matter, and there might still be a single long-distance monopoly. But no such inquiry was undertaken before MCI's offering of Execunet service, and the courts decided that the FCC had not established a sufficient record to bar competition in switched long-distance services.

In 1978 the commission launched an inquiry into the desirability of competition in switched services, but the outcome of this inquiry was largely foreordained.[18] Once the FCC and the Department of Commerce persuaded AT&T and the new specialized carriers to negotiate a temporary access tariff, the new carriers gained de facto acceptance. It would have been virtually impossible to conclude the investigation in 1980 or 1981 with an order banishing the new carriers from the market for switched services.

16. Based on data published by the U.S. Telephone Association and the Federal Communications Commission, *Statistics of Communications Common Carriers* (Washington, annual issues).

17. *MCI Telecommunications Corp.* v. *Federal Communications Commission*, 561 F. 2d 365 (D.C. Cir. 1977).

18. *Further Notice of Proposed Rulemaking*, FCC CC Docket 78-72. *Federal Register*, vol. 49, April 30, 1978, pp. 18, 318.

Distortions in Telephone Ratemaking

For most of the period between 1934 and 1960, the FCC was not forced to regulate individual AT&T rates. As long as AT&T's rate of return did not rise above a threshold level, its rates would be adjudged "just and reasonable." Implicitly, AT&T had to maintain switched long-distance rates that were geographically uniform, regardless of the traffic density on the route. Undoubtedly, such a standard was economically inefficient but politically reasonable. Some services might be more profitable than others, and traffic on dense routes may have been subsidizing the traffic on routes with less traffic. As long as users did not complain, however, the FCC was unlikely to intervene.[19] The average rates had to be just and reasonable, but individual rates did not have to reflect costs.

Regulated rates for telephone service reflected the political pressures that the Federal Communications Commission and state regulators faced. Though quasi-efficient pricing in a market with rapid technical change and a predominance of joint and common costs may have been difficult, the departures from efficiency that developed through the regulatory process reflected more than the technical problems in measuring costs.

Over time, at least five types of rate distortions developed in this regulated sector:

—Long-distance rates were held artificially high so as to mitigate increases in local rates.

—Long-distance rates were based on distance, not on call density.

—Local rates were usually lower for high-cost rural areas than for lower-cost urban areas.

—Business users were charged more for access and local exchange service than residential users in the same exchange.

—Local service was generally offered on a flat-rate basis; therefore, heavy users paid no more for local service than users who placed less of a burden on traffic-sensitive facilities.

Each of these distortions invited some form of competitive entry if the regulators could be persuaded to allow it. Indeed, some entry

19. Some users did complain. For example, the commercial television networks often argued that AT&T's rates for video interconnection were too high.

subsequently occurred, often without the regulators' knowledge or acquiescence.

Cost Recovery—Local versus Long-Distance Rates

Even if the FCC did not closely regulate individual rates, federal and state regulators were forced to allocate costs between federal (interstate) and state jurisdictions. Because local, intrastate toll, and interstate toll calls are originated and terminated on the same local subscriber loops and switched by the same local switches, the costs of these facilities must be allocated between the two jurisdictions. Rates in either jurisdiction vary directly with the share of costs allocated to that jurisdiction.

Until World War II, the cost of local loops was borne fully by local exchange companies. This pricing principle was challenged in *Smith* v. *Illinois Bell,* when the Supreme Court ruled that Illinois Bell could not assign all of the local loop costs to local service.[20] Responding in part to this Court decision, in 1943, a joint board of federal and state regulators decided to allocate local plant costs between local and interstate jurisdictions in proportion to the minutes of use in each category. Since interstate use constituted only 3 percent of total minutes of use at that time, this change in policy constituted only a minor repricing of service.[21]

As the share of interstate use of the network rose, the share of the non-traffic-sensitive costs allocated to interstate use also rose, but this change did not satisfy state regulators.[22] In the 1950s, as microwave transmission began to lower long-distance transmission costs, regulators introduced a new formula to allocate a larger share of local subscriber plant costs to the interstate service, thus allowing local rates to be depressed relative to costs, keeping long-distance rates artificially high, and supporting high-cost, local telephone companies.

The formula—the subscriber plant factor—used in allocating the

20. *Smith* v. *Illinois Bell Telephone Co.*, 282 U.S. at 151 (1930).

21. For a useful summary of these policies, see Peyton L. Wynns, *The Changing Telephone Industry: Access Charges, Universal Service, and Local Rates* (Washington: Congressional Budget Office, June 1984).

22. Non-traffic-sensitive costs (NTS) are those costs that do not vary with the intensity of traffic. The most important of these NTS costs are those associated with subscriber connections.

Figure 2-1. Percent of Subscriber Plant Costs Allocated to Interstate Service

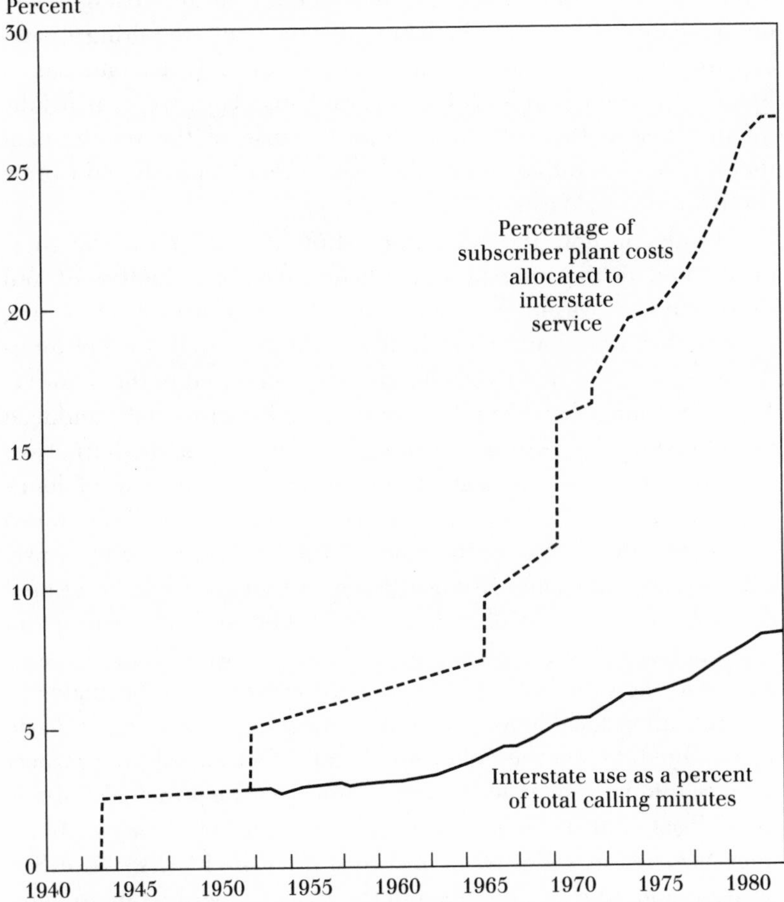

Source: Congressional Budget Office, *The Changing Telephone Industry: Access Charges, Universal Service, and Local Rates* (Washington, June 1984), p. 10.

local loop costs was changed twice more, in 1965 and 1969, in each case raising the interstate share of local costs. By 1981 the interstate share of usage had risen to 8 percent, but regulators were allocating 27 percent of non-traffic-sensitive costs to interstate calls (figure 2-1).

To many observers, the shift of costs from the intrastate to the interstate jurisdiction constituted a subsidy from long-distance to local service. Strictly speaking, for this cost-shifting to represent a subsidy, it would have to result in the price of interstate service being above the stand-alone costs of providing this service, including the costs of providing local connections for interexchange calls alone. There is also a question about whether the local exchanges are designed explicitly for handling long-distance calls, requiring greater expense than for exchanges that would handle only local messages. Some even argue that AT&T's local exchange costs were artificially inflated before the 1984 divestiture because of the requirement that the Bell operating companies' equipment be purchased from a captive supplier, Western Electric.

Subsidy or not, the regulatory shift of costs from the local jurisdiction to the interstate jurisdiction was both inefficient and shortsighted. It was inefficient because the fixed costs of connecting a subscriber were partially reflected in the marginal price of long-distance calls, not in a fixed charge to be connected to the network. The higher long-distance rates that resulted from this policy induced users to reduce long-distance calling or, later, to look for alternatives to common-carrier message telephone service. The use of long-distance calls to pay subscribers' fixed connection costs reduced economic welfare because these non-traffic-sensitive costs obviously did not vary with long-distance usage. Although it can be argued that some subsidy to local access is efficient because of consumption externalities in communications markets, the use of long-distance service as the vehicle to cross subsidize access was unfortunate.[23]

The shift of local loop costs to the interstate jurisdiction was also shortsighted. When the policy was begun, there was little prospect of entry into long-distance services. However, the principal shifts of these local non-traffic-sensitive costs to interstate long-distance rates occurred in 1965, 1969, and 1971 after the FCC's Above-890 decision and after MCI applied to the FCC for permission to enter as a specialized carrier. The regulatory decision to raise long-distance rates relative to local usage and access rates simply increased the probability of entry into long-distance services.

During the late 1960s, the FCC began to look more favorably on entry into interstate services, but the state regulators and AT&T

23. See Jeffrey Rohlfs, "A Theory of Interdependent Demand for a Communications Service," *Bell Journal of Economics*, vol. 5 (Spring 1974), pp. 16–37. 6. See also John T. Wenders, *The Economics of Telecommunications: Theory and Policy* (Ballinger, 1987), chap. 4, for a discussion of optimal pricing of telephone service in the presence of externalities and a revenue constraint.

obviously had a different view. Why these state regulators should have pressed for an artificial increase in long-distance rates just as the floodgates of entry were opening is unclear. Even more puzzling, however, is AT&T's apparent acquiescence in this distortion in ratemaking. At that time its local services were largely free from any possible threat of entry, but its long-distance business was coming under attack from new entrants.[24]

Uniform Geographic Rates

It is not difficult to understand why politically responsible regulators would want AT&T rates to be uniform across all market segments of equal length.[25] If rates for two different 1,000-mile calls were not the same, customers would feel that the high-rate route was being unfairly penalized even though the cost per call on this route was much higher than the cost on denser routes. This reality is simply one example of the populist appeal that creates rural subsidies.

Unfortunately, geographically uniform prices created artificial inducements to entry into long-distance communications after 1959. If rates were deliberately kept above costs on dense routes so that similar rates could be charged on the less dense routes, new competitors would find the former attractive and ignore the latter. This is, of course, exactly what happened in the 1970s as MCI began to offer service between large cities, at first ignoring the underpriced smaller markets.

AT&T responded to MCI's entry by offering a "hi-lo" tariff that reduced rates to larger communities.[26] When this private-line tariff was disallowed, it filed a new tariff—the multischedule, private-line (MPL) tariff—that also differentiated between low-capacity and high-capacity facilities. This multischedule tariff was also challenged, but after more than three years of FCC proceedings, the

24. The usual explanation for AT&T's acquiescence is that it preferred to keep long-distance rates from falling rather than asking state regulators for local subscriber rate increases.

25. Chuck Jackson has reminded me that geographic uniformity does not exist across jurisdictional boundaries. A one-hundred-mile call may be priced quite differently in California and New York, for example. In the federal jurisdiction, however, geographic uniformity has been the implicit goal of the FCC.

26. *American Telephone and Telegraph*, 58 FCC 2d 362 (1976).

tariff was essentially allowed to remain in effect when the FCC admitted that it could not measure AT&T's individual service costs.[27] By insisting on the impossible, a definitive allocation of joint and common costs, the FCC only harassed AT&T and encouraged uneconomic entry.

Rural-Urban Distortions In Local Rates

State commissions are responsible for the regulation of local access and exchange rates. These politically responsive institutions usually find the pressures for rural subsidies irresistible. In telephony, the cost of service declines rapidly with population density because the length of the average subscriber loop is shorter in dense areas. If rates were set to reflect costs, rates in urban areas would be far below those in rural areas.[28] In fact, the opposite pattern has persisted: urban rates have been above rural rates.

Business Rates versus Residential Rates

If populist pressures have led to distorted relationships between urban and rural rates, they have surely been responsible for the fact that single-line and multiline business rates are often far above residential rates even though the costs of the two types of service are nearly identical. Businesses tend to cluster closer to the central-office switch, leading to lower average loop costs, but they use their lines somewhat more intensively. The resulting costs for typical flat-rate service are likely to be similar for residences and businesses,[29] but rates have usually been much higher for businesses than for residential subscribers.

Flat-Rate versus Measured Service

Until recently, most local service in the United States was offered on a flat monthly basis within a specified local calling area. This

27. *American Telephone and Telegraph*, 74 FCC 2d 1 (1979), *recon.*, 85 FCC 2d 549 (1981).

28. See Bridger M. Mitchell, "Incremental Capital Costs of Telephone Access and Local Use," prepared for the Incremental Cost Task Force R–3764–RC (Santa Monica, Calif.: Rand Corporation, August 1989).

29. Mitchell, "Incremental Capital Costs."

pricing policy allowed subscribers to use the local switching facilities at a zero price even though the marginal cost of using the network is above zero. Attempts to move to measured service have traditionally met great political opposition even though such pricing may lead to a more efficient (lower-cost) system of access.[30] Nevertheless, the phasing down of the subsidies from interstate services and the increase in the bypass of telephone-company circuits have forced state regulators to begin to approve measured-service rate plans.

Regulatory Responses to Competition

When the FCC began to allow competition in a limited array of interstate services, it invited an attack on the rate distortions that it had helped to create. Very soon, the commission had to face the necessity of setting new rules for AT&T's response, for AT&T would obviously wish to respond to this assault on its erstwhile monopoly by greatly altering its rate structure. Eventually, the FCC would be forced into an overhaul of its entire approach to interstate rate regulation.

Regulating Individual Rates

At first, competitive entry was limited to interstate services. The new competitors in private-line and (later) switched-message services did not simply attack AT&T along dense routes. They knew that large customers account for a very large share of long-distance service. These large customers, AT&T realized, would be the principal targets for the new entrants, and it therefore responded by offering large discounts to large users.

These discounts began even before MCI entered in 1969. In 1961 AT&T submitted its first Telpak tariff for private-line customers. It offered very large discounts to large customers to dissuade them from investing in private microwave. The FCC disallowed the

30. Recent research by Rolla Edward Park and Bridger M. Mitchell, "Optimal Peak-load Pricing for Local Telephone Calls," R–3404–1–RC (Santa Monica, Calif.: Rand Corporation, March 1987), concludes that the welfare gains from feasible usage-sensitive pricing are virtually nonexistent.

discounts for the smallest classes of customers, but the commission could not decide on the other discounts for the next fifteen years. An investigation into these rates became mired down in problems of cost-accounting methodology and the inadequacy of data. As with the hi-lo and multischedule, private-line tariffs, the FCC found it could not meaningfully allocate costs across services or customers. Such an exercise was meaningless anyway, some economists argued, because rates should not be determined by accounting allocations of historical costs.[31] Given the predominance of joint and common costs, the rapid rate of technical progress, and possible increasing returns to scale, any rate determined by these historical allocations is unlikely to approximate an efficient rate.

In the 1960s, AT&T also began to offer WATS (wide-area telephone service), a discounted tariff offering of voice-grade, long-distance service. This discounted service fended off competition and prevented some flight to private lines, but its effectiveness was severely curtailed once the FCC began to require AT&T to allow resale of WATS services by arbitragers to smaller customers.

The FCC tried to measure the costs of each broad service category through a cost-allocation process until 1989, but it simply could not estimate the cost of every service option offered by AT&T. In effect, the commission had to approve or disapprove individual rates based on its perception of the "fairness" of such rates. AT&T's competitors, whose rates are totally unregulated, predictably have opposed most AT&T rate reductions, but the commission has increasingly allowed such reductions even if it cannot know whether these rates are above or below an appropriate measure of cost.[32]

Repricing Long-Distance and Local Service

In the late 1970s, the FCC began to recognize the inefficiency of allowing non-traffic-sensitive costs to be recovered from long-

31. AT&T repeatedly made this argument. See *Competition and Monopoly In Telecommunications Services*, attachment to Bell Exhibit 16, appendix B, FCC CC Docket 19919, November 23, 1970. For a useful discussion of this issue, see Alfred E. Kahn, *The Economics of Regulation: Principles and Institutions*, vol. 1 (Wiley, 1970), pp. 151–58.

32. The latest of these AT&T tariffs, tariff 15, is being offered to large customers (such as Holiday Inns) in order to compete with the unregulated carriers who are free to offer any discount they choose. In addition, the FCC has allowed AT&T to respond with discounted ordinary switched services through its tariff 12.

distance calls.[33] It therefore proposed to phase out non-traffic-sensitive cost recovery from access charges on long-distance carriers by instituting fixed monthly subscriber-line charges (SLCs) on business and residential customers. By reducing carrier-access charges, the commission would dissuade the long-distance carriers from seeking alternative technologies to bypass the local telephone companies in originating and terminating their calls. Unfortunately, strong political opposition has frustrated the full implementation of this plan.[34]

The FCC has also used the subscriber plant factor (SPF) to limit the degree to which companies may allocate the fixed costs of their subscriber loops to the interstate jurisdiction.[35] The SPF is the ratio of non-traffic-sensitive plant costs allocated to interstate long-distance calls. The SPF rose steadily over time for most companies, allowing them to shift a larger share of their costs to the interstate jurisdiction. By 1981 the commission had assigned 27 percent of non-traffic-sensitive costs of subscriber loops to interstate long-distance services, but only 8 percent of total usage was attributable to interstate calls. The FCC has frozen SPF and will reduce it to 25 percent within eight years.

Each of these decisions reflected the FCC's aim to induce more efficient (that is, cost-based) pricing of telephone service. Because the cost of local subscriber loops does not vary with the use of these circuits, defraying this cost from long-distance charges is inefficient. As the FCC decisions reduced the local carriers' ability to shift costs from local to interstate service, they raised other rates that are within the state regulators' purview.

The effect of the repricing of access to the network may be seen in table 2-1. The average cost of premium interstate access per minute of conversation paid by all interexchange carriers for customer connections in exchanges offering equal access has fallen greatly over the past five years.[36] As subscriber-line charges have risen, the cost of premium access for interstate long-distance calls has fallen by more than 40 percent between 1984 and 1989. By

33. FCC CC Docket 78-72.

34. See chap. 5.

35. Amendment of Part 67 of the Commission's Rules, 55 RR 2d 659 (1984).

36. "Premium" refers to high-quality access provided to all carriers in exchanges equipped with equal-access capability. Otherwise, it applies only to access charges for AT&T because AT&T is the only carrier with full, high-quality services in these latter exchanges.

Table 2-1. The Repricing of Access to the Telephone Network

	Monthly subscriber-line charges		
Date	*Residential and single-line business (dollars per line)*	*Multiline business (dollars per line)*	*Average "premium" interstate access charge per minute of conversation (dollars per minute of conversation)*
1/1/85 to 5/31/85	0.00	4.99	0.177
6/1/85 to 9/30/85	1.00	4.99	0.162
10/1/85 to 5/31/86	1.00	4.97	0.154
6/1/86 to 12/31/86	2.00	4.97	0.140
1/1/87 to 6/30/87	2.00	5.12	0.124
7/1/87 to 12/31/87	2.60	5.12	0.115
1/1/88 to 11/30/88	2.60	5.01	0.106
12/1/88 to 3/31/89	3.20	5.01	0.096
4/1/89 to 12/31/89	3.50	4.94	0.095

Source: "Monitoring Report of the Federal-State Joint Board," in Federal Communications Commission CC Docket 87-339, December 1988, pp. 328–29, and January 1990, pp. 246–47.

the middle of 1989, subscriber-line charges for residential customers were $3.50 a month while those for multiline businesses were still capped at a maximum of $6.00 a month.[37]

This repricing of telephone service is far from complete if the goal is to shift the cost of supporting the non-traffic-sensitive subscriber costs from interexchange access charges to fixed monthly subscriber line charges. In 1984 total interstate access charges were $14.4 billion while subscriber-line charges accounted for only $681 million in total carrier revenues.[38] By 1988 reported revenues from subscriber-line charges had risen to $4.3 billion, but interstate

37. The average multiline business subscriber-line charge in table 2-1 reflects that the federal share of NTS costs per line is much less than $6 a month in many states.

38. FCC, *Statistics of Communications Common Carriers* (1988–89), p. 240.

Table 2-2. Access Charges, End-User Fees, and Total Revenues, Local-Exchange Carriers, 1984, 1988
Millions of current dollars

Charges and revenues	*1984*	*1988*
Interstate access charges (excluding end-user charges)	14,384	15,128
Intrastate access charges	5,335	6,051
Total access charges	19,719[a]	21,180[a]
Total operating revenues	69,597	81,171
Access charges/revenues	0.283	0.261
End-user charges (subscriber-line charges)	681	4,327
End-user charges/revenues	0.010	0.053

Source: FCC, *Statistics of Communications Common Carriers* (Washington, 1984, 1988–89), p. 24; and pp. 36, 240, respectively.
a. Total access charges are the total network access revenues minus the end-user revenues.

access charges still generated $15.1 billion in local carrier revenues (table 2-2). Even with the final $3.50 monthly residential subscriber-line charge, the total annual yield from all such charges will rise to only about $5.1 billion, or about two-thirds of the interstate assignment of non-traffic-sensitive costs.[39] Of course, three-fourths of all non-traffic-sensitive costs must be recovered from local and intrastate services over which the FCC has no jurisdiction.

Competition in Equipment

Competition in terminal equipment developed more deliberately, but not without AT&T opposition. As late as the 1960s, attaching "foreign equipment" to a telephone line violated the tariffs of regulated carriers. This prohibition extended from such elementary devices as rubber cups attached to the telephone handset to more complicated devices that interconnect telephone lines to other communications media. All telephone sets, private branch exchanges, and other standard equipment used by residences or businesses were owned and leased by the telephone company.

Virtually all of AT&T's customer premises equipment (CPE) was manufactured by Western Electric and sold to the operating

39. Even if all NTS costs now assigned to the interstate jurisdiction were eliminated, there would still be substantial interstate access charges because of traffic-sensitive costs that represent about one-third of the cost basis for 1990 interstate switched access charges.

companies at prices not subject to a competitive bid. The Bell operating companies had no choice, but they did not have to fear competition. In nearly every community, only one regulated telephone carrier provided service, including the services of the terminal equipment.

Between 1968 and 1978, the Federal Communications Commission gradually ended the carriers' monopoly over terminal equipment. AT&T, its local companies, and even state regulators had argued strenuously for a protective device to be installed between foreign equipment and the telephone line, but the FCC would not allow it. In the end, the FCC simply required that all equipment be certified as safe for use on the network.[40]

Regulatory Shrinkage of the Rate Base

The liberalization of the equipment market led to a shift toward consumer-owned terminal equipment. But it also raised interesting questions about the regulated carriers' leasing of this equipment to its customers. Why should this activity be regulated at all? This concern, in turn, led to an inquiry into whether customer premises equipment and even the wires inside a customer's home or establishment should be in the rate base. In 1980 the FCC initially decided that it could fully deregulate CPE only if the regulated carriers who sold or leased such equipment would establish separate subsidiaries.[41] This rule would be changed in the wake of the 1984 AT&T divestiture.

Competition in terminal equipment developed slowly until the late 1970s. In the 1980s, however, regulatory-induced changes in CPE and inside wiring began to seriously affect the rate base—the embedded investment in plant and equipment required to serve the market—from which local charges could be extracted.

The first of these changes was the expensing of inside wiring. In 1981 the FCC required a phasing in of the requirement that wiring inside a customer's premises be expensed rather than capitalized and placed in the rate base.[42] This requirement forced telephone companies to charge the customer directly for installation rather

40. See chap. 4.

41. *Second Computer Inquiry*, 77 FCC 2d 384 (1980).

42. *Second Computer Inquiry*, 85 FCC 2d 818 (1981).

than adding his inside wiring to the rate base and recapturing part of the cost from long-distance rates through the process of separations and settlements.

In 1982 the FCC asserted control over the telephone companies' depreciation policies. In this decision, the commission sought to accelerate the depreciation of certain categories of equipment so as to reduce the tendency of obsolete equipment to accumulate in the rate base and to increase the incentives for technological change.[43] The courts subsequently overturned this decision.[44]

Next the FCC required that new CPE be detariffed beginning in 1983 and that embedded equipment be detariffed by the Bell companies by January 1, 1984.[45] This ruling preceded the effective date of the AT&T divestiture, which required that existing CPE be transferred from the operating companies to AT&T. This decision removed CPE from the Bell operating companies' rate base, but the FCC's decision also required non-Bell companies to phase out their CPE from their rate bases by a later date. The Bell operating companies can now sell (but not manufacture) CPE, but the cost of such sales are not supposed to be included in the costs of regulated, local access/exchange services.

Antitrust - II

For sixty years, AT&T was able to avoid the strictures of antitrust. In 1913 and 1956, it held off threatened suits or settled an actual antitrust suit by agreeing to limitations on its activities. In the 1970s, however, AT&T began to find its political power insufficient to hold off the regulatory reformers and the trustbusters. First, AT&T lost almost every battle over competitive entry at the FCC and in the courts. Second, AT&T failed in attempts to pass legislation that would have insulated it from competition in interexchange services and terminal equipment. Finally, it could not prevent the filing of an antitrust suit in 1974—this time by a Republican administration.

In October 1974 the Justice Department filed the suit that

43. *Uniform System of Accounts*, 89 FCC 2d 1094 (1982).

44. *Louisiana PSC* v. *Federal Communication Commission*, 476 U.S. 355 (1986).

45. *Second Computer Inquiry*, 95 FCC 2d 1276 (1983). *Second Computer Inquiry*, 98 FCC 2d 814 (1984). Detariffing means that all customer premises equipment (CPE) had to be offered for sale or rental separately and that such equipment could not be placed in the rate base for setting local or long-distance rates.

eventually resulted in the dismemberment of AT&T. It charged that AT&T had violated Section 2 of the Sherman Act by monopolizing interstate communications services and the market for telecommunications equipment.[46] The initial complaint sought the divestiture of Western Electric and Bell Laboratories from AT&T.

The 1974 suit occurred before the courts ruled on Execunet and terminal equipment issues but after specialized carriers started competing in earnest with AT&T for private-line business. MCI, in particular, had begun to ask for a variety of AT&T operating company circuits to originate and terminate its calls. Undoubtedly the frustration of the new long-distance competitors—particularly MCI—who were trying to compete with AT&T while being customers of AT&T for local circuits eventually persuaded the government to file the antitrust suit. Probably a suit would not have been brought on equipment issues alone. Had AT&T been more accommodating in allowing access to its local circuits to rival interexchange carriers, it might have survived into the 1980s as an integrated long-distance and local-exchange carrier. It chose a more aggressive path, however, and this behavior sealed its doom in the antitrust courts.

The 1974 suit has at least two ironical aspects. First, the Justice Department suit was brought against a regulated company that had been under FCC scrutiny for forty years. As the suit was being filed, the FCC—the government's expert telecommunications agency—was investigating the structure of AT&T and eventually decided to recommend few changes.[47] The Justice Department, far less expert on these matters, essentially overruled this judgment and was vindicated in the 1980s.

Second, the suit was brought in the waning days of a Republican administration that was generally hostile to new antitrust actions under Section 2 of the Sherman Act and Section 7 of the Clayton Act. Antitrust had reached its recent zenith in the Johnson administration, with the filing of the IBM suit and the publication of the Neal Commission report. The Nixon and Ford administrations

46. *United States* v. *American Telephone and Telegraph Co.*, Civil Action 74-1698 (D.D.C. November 20, 1974). Specifically, the complaint charged that AT&T had abused its power as a bottleneck monopolist in procuring equipment, excluding competition from the terminal equipment market, and denying access to long-distance competitors.

47. *Charges for Interstate Telephone Service*, 64 FCC 2d I (1977).

retreated from these bold strategies aimed at reducing industrial concentration. The AT&T case was thus an exception, apparently a case brought with little consultation of the political authorities.[48] The Reagan Justice Department eventually abandoned the IBM case, but the AT&T case was concluded by a Justice Department in which the top two politically responsible officials were forced to recuse themselves from the case.

The 1982 Decree

The AT&T case was destined to become one of the most extensive and expensive antitrust litigations in history. From the outset, a struggle took place in the Antitrust Division over the resources required to prosecute the case. The first lead attorney resigned after being denied the resources he felt were necessary. The second person in this position resigned before the case came to trial. The first judge to whom the case was assigned died and was replaced by Judge Harold Greene, who would see the case to a conclusion.[49]

AT&T first moved to have the case dismissed on jurisdictional grounds. After all, AT&T had been a regulated company for more than half a century. This motion for dismissal was denied, thereby forcing a trial on the merits of the Justice Department's case.

After nearly six years of preliminary skirmishing, the Justice Department finally began to place its case before the court. By the fall of 1981, the government's case had been concluded, and AT&T moved for summary judgment. Judge Greene denied this motion in such strong terms that it sent shock waves through the defense.[50] The government, Greene asserted, had presented a strong case that created at least the rebuttable presumption of violations of the Sherman Act. Greene concluded that the government had presented substantial evidence of AT&T's denial of access to MCI and other

48. For a discussion of the history of the case, see Steve Coll, *The Deal of the Century: The Breakup of AT&T* (Atheneum, 1986).

49. Some might argue that the case has still not been concluded because Greene retains jurisdiction over the 1982 decree that settled it.

50. *United States* v. *American Telephone and Telegraph Co.* 524 F. Supp. 1336 (1981).

carriers, aggressive pricing in response to entry, and AT&T's use of Western Electric to frustrate independent equipment manufacturers in trying to sell to Bell operating companies or AT&T Long Lines. Of course, AT&T still had the opportunity to present its defense, but clearly Greene believed that AT&T would have little prospect for victory at the district court level.

The change in administrations in January 1981 might have been the opportune time for AT&T to seek a political settlement of the case, just as it had in the 1953–56 period for the earlier case. This time, however, the attorney general and deputy attorney general had to recuse themselves from the case. The assistant attorney general, William Baxter, was therefore left to carry the department's position to Cabinet-level discussions on this matter. He left no doubt that he wanted "to litigate the case to the eyeballs," and he successfully parried opposition from the Commerce and Defense Departments. By the time the opposition had time to coalesce, the government's case had begun. Baxter could not be stopped without risking a confrontation with Judge Greene and a public-relations problem.

After Judge Greene dismissed AT&T's motion for summary judgment, settlement talks began in earnest. The Justice Department insisted that the chief source of antitrust liability was AT&T's ownership of the "bottleneck" facilities of the Bell operating companies. The ownership of these bottlenecks provided AT&T with the leverage to exclude competition in interexchange (long-distance) and other services requiring the local companies' circuits to originate or complete a call. And these operating companies could also exclude competitors of Western Electric from selling CPE and the capital equipment—switching and transmission equipment—used by AT&T in offering local services.

After several months of deliberation, the Justice Department and AT&T announced their agreement to settle the case. AT&T would divest itself of all operating companies but retain its Western Electric and Long Lines divisions. Bell Laboratories would also remain with AT&T, but some of its personnel would be transferred to a new research organization for the divested Bell operating companies, to be called Bellcore. The newly independent operating companies would be allowed to provide only "local" services so as to prevent the reassembly of the vertical monopoly that provoked the case in the first place. Finally, AT&T was to be freed of the restrictions on its activities that were built into the 1956 decree, thereby permitting it to enter other electronics businesses, including computers.

Many details had to be ironed out before the decree could be entered with finality before the court. Under the Tunney Act, the decree had to be reviewed by the court and incorporated as a "modification" of the 1956 decree, resulting in its being called the modified final judgment or simply the MFJ.[51]

Among the provisions in the MFJ, as finally approved by Judge Greene, were explicit requirements that the newly divested Bell operating companies move quickly to install switches that would allow equal access for all interexchange carriers.[52] AT&T had to agree to continue to supply capital equipment to these companies at reasonable prices. The divested Bell operating companies were forbidden to engage in competitive telecommunications businesses, such as equipment manufacturing, long-distance service outside of new local access and transport areas (LATAs), and information services. In a curious decision, however, Greene forced the Justice Department to allow the divested Bell operating companies to keep their Yellow Pages. He also allowed the Bell operating companies to sell (but not to manufacture) CPE. And in a concession to the powerful newspaper industry, Greene required AT&T to refrain from electronic publishing over its own lines for five years.

In the next two years, a new telephone industry was carved out of the old giant AT&T under a plan of reorganization, approved by Judge Greene. The market for communications equipment and long-distance services would clearly change. The Bell operating companies were left to squeeze whatever profits they could from the one monopoly business left—local service—subject, of course, to state regulation. It was widely believed that AT&T had triumphed, shedding itself of the tired old business of local exchange services

51. Modification of Final Judgment, *United States* v. *American Telephone and Telegraph*, 552 F. Supp. 131 (D.D.C. 1982), *aff'd. sub. nom.*, *Maryland* v. *United States*, 460 U.S. 1001, 103 S. Ct. 1240, 75 L. Ed. 2d 472 (1983).

52. As of the fourth quarter of 1989 an estimated 94.9 percent of all Bell operating company lines and 81.6 of independent telephone company lines had been converted to equal access. Common Carrier Bureau, Industry Analysis Division, *Trends in Telephone Service* (Federal Communications Commission, February, 1990), p. 14.

while being freed to enter the dynamic world of computers.[53] Clearly, it was thought, the divested Bell companies would begin to wither but retain the local access/exchange business until new technology made this service competitive as well. Otherwise, competition was to replace FCC regulation, and Judge Greene was to share responsibility with the FCC for regulating interstate telephone markets.

Regulation in the Postdivestiture World

Since 1984 little further progress toward deregulation has occurred. Only terminal equipment is fully deregulated; all common-carrier services remain under federal and state regulation.

At the interstate level, AT&T remains a regulated, "dominant" carrier, whose tariffs must be approved by the FCC. Competitive interexchange carriers—the OCCs—are not subject to rate regulation, however, because the commission has determined that competitive forces are sufficient to ensure that their rates are "just, reasonable, and non-discriminatory."[54] AT&T's pricing responses to the OCCs' offerings are thus subject to challenge from the OCCs. But because the commission cannot accurately measure AT&T's individual service costs, these disputes are not easily resolved.

In the past three years, the FCC has been implementing a new approach to regulation of dominant carriers involving price caps rather than rate-of-return regulation.[55] In 1989 the commission adopted price caps for AT&T and proposed a price-cap plan for the local exchange carriers' interstate activities.[56] More recently, the FCC has proposed essentially to deregulate AT&T's business

53. See, for example, Paul W. MacAvoy and Kenneth Robinson, "Winning by Losing: The AT&T Settlement and Its Impact on Telecommunications," *Yale Journal on Regulation*, vol. 1, no. 1 (1983), pp. 1–42.

54. These policies have been decided in the FCC's *Competitive Carrier* proceedings. See *Policy and Rules concerning Rates for Competitive Common Carrier Services and Facilities Authorizations*, 77 FCC 2d 308 (1979). A series of decisions ensued between 1980 and 1985, the last being the *Sixth Report and Order*, 99 FCC 2d 1020 (1985).

55. *Policy and Rules concerning Rates for Dominant Carriers* FCC CC Docket 87–313.

56. *Policy and Rules concerning Rates for Dominant Carriers*, FCC CC Docket 87-313, adopted March 16, 1989, released April 17, 1989.

services—a proposal that has predictably generated serious opposition from the OCCs.

For "enhanced" telecommunications services, FCC rules require that AT&T and the regional Bell operating companies provide open network architecture or comparably efficient interconnection to ensure that competitive service vendors have access to bottleneck-facility services on the same terms as those offered by the carriers to themselves.[57] The goal of this requirement is to encourage transparency in carriers' access arrangements.

At the state level, most intrastate toll and local exchange service remains tightly regulated. Some states have liberalized entry into these markets, and some have even begun to experiment with new regulatory approaches, such as price caps. But only Nebraska has totally deregulated intrastate telecommunications.

Summary

Despite the popular belief that the telephone network is a natural monopoly, the AT&T monopoly survived until the 1980s not because of its naturalness but because of overt government policy. In the late 1950s, technological change, regulatory distortions, and entrepreneurial energy combined to create pressures for entry into various facets of the telecommunications industry. The expert agency entrusted with federal telephone regulation, the FCC, found it increasingly difficult to deny entrants the right to offer equipment and services in direct competition with AT&T. The FCC lost control of the process, however, because it could not successfully arbitrate disputes between the incumbent (AT&T) and the new challengers. When the new competitors failed to get what they wanted from the FCC, they changed their strategy and looked for relief in the antitrust courts. In 1974 these competitors persuaded the Justice Department to file a mammoth antitrust suit against AT&T, and the stage was set for the last battle of AT&T, which AT&T lost.

AT&T was dismembered on January 1, 1984. The United States is now seven years into an experiment in vertical fragmentation of a crucial industry that accounts for more than 2 percent of gross

57. *Third Computer Inquiry*, 104 FCC 2d 958 (1986), *clarified on recon.*, 2 FCC Rcd 3035 (1987).

national product. The following chapters examine how competition and divestiture have affected telephone service, productivity, rates, equipment prices, and even employee wages. Competition and divestiture have not replaced regulation, however, for most federal and state regulation of rates remains. Unfortunately, this rather dangerous combination of regulation and competition creates many of the tensions in today's telecommunications environment.

Chapter Three

Telephone Services and Rates

TECHNOLOGICAL CHANGE, shifts in demand, the advent of competition, and divestiture have transformed the U.S. telecommunications services sector, placing great pressure on regulators, who have responded by changing the structure of telephone rates. These rate changes, in turn, have affected the relative growth rates of local carriers and long-distance companies. In addition, the pressures of competition and divestiture have affected the relative efficiency of all carriers. In this chapter, I analyze the impacts of divestiture and entry liberalization on telephone rates, the growth in industry output, and productivity growth. The effects of these changes on economic welfare are addressed in chapter 5.

The Growth of the Telephone Services Sector

To most observers, the telephone industry is composed of those firms regulated by the Federal Communications Commission (FCC) and state public utility commissions. They include local exchange companies, interexchange carriers (long-distance companies), and telegraph companies (principally Western Union). All of the larger carriers report the results of their operations annually to the FCC. Many other suppliers of telephone services do not report to the FCC even though they may need FCC authority to construct their facilities. Most of these suppliers are private carriers or private networks that offer their services principally to their parent corporations. Discussions about telephone industry policy often overlook the growth in such operations.

The services of the regulated common carriers fit into the conventional categories of local and long-distance services. The

Table 3-1. The Real Output of Telecommunications Services, 1972–88
1977 = 100

Year	*Local*	*MTS*[a] *(inter-state)*	*MTS*[a] *(intra-state)*	*WATS (inter-state)*	*WATS (intra-state)*	*Total output*[b]
1972	71.7	71.3	67.6	34.4	40.9	69.3
1973	77.1	79.1	75.9	43.2	50.7	76.1
1974	81.7	85.6	82.3	52.9	60.4	81.7
1975	86.2	87.5	86.8	63.6	68.5	85.6
1976	91.3	91.9	92.4	81.0	82.8	91.2
1977	100.0	100.0	100.0	100.0	100.0	100.0
1978	106.6	113.4	115.2	122.5	128.4	111.0
1979	114.1	130.9	129.3	143.2	148.8	123.2
1980	122.6	145.7	141.7	165.5	170.6	134.9
1981	125.9	157.0	150.7	190.9	191.9	142.5
1982	129.7	154.8	154.9	227.8	227.1	146.5
1983	128.5	155.3	155.0	282.7	288.8	149.9
1984	104.7	167.0	171.1	352.7	424.0	149.3
1985	104.2	185.6	171.4	404.7	487.7	156.6
1986	104.5	210.8	184.9	464.7	541.4	167.9
1987	104.7	257.9	204.2	479.7	564.2	183.8
1988	106.0	284.9	225.9	489.5	529.1	194.6
			Average annual percent change			
1972–83	5.3	7.1	7.5	19.1	17.8	7.0
1983–88	−3.8	12.1	7.5	11.0	12.1	5.2

Sources: Revenue figures for each service based on author's calculations from Federal Communications Commission (FCC), *Statistics of Communications Common Carriers* (Washington, annual issues); United States Telephone Association (USTA), *Independent Telephone Statistics*, or *Telephone Statistics*, vol. 1 (Washington, annual issues), and *Statistics of the Local Exchange Carriers, 1989* (Washington, 1989); Laurits R. Christensen, Dianne C. Christensen, and Philip E. Schoech, "Total Factor Productivity in the Bell System, 1947–79," Christensen Associates, Madison, Wis., September 1981, p. 13. The division of intra- and interstate message telecommunications service (MTS) and wide area telephone service (WATS) is based on AT&T Long Lines Business Research. "Selected Interstate Data under the Division of Revenues Contracts and Long Lines Statistics, 1960–1982," Murray Hill, N.J., April 1983, which provides a breakdown of AT&T's MTS and WATS revenues. Local service reflects the detariffing of terminal equipment in 1984. All series are deflated revenues, using the Bureau of Labor Statistics producer price indexes as deflators.

a. Includes private lines and other toll.

b. Total deflated revenues (not a Divisia index of the subcategories).

most important of the long-distance services are switched-message telephone services (MTS), private-line services, and WATS, a discounted bulk offering of switched services. The trends in the output of each service are measured in table 3-1 by deflating the total revenues from each, as recorded by all reporting carriers, by the appropriate producer price index.[1]

1. The producer price index measures the prices received by telephone companies for services rendered. It therefore excludes payments by consumers for those services offered by non-telco suppliers.

The data in table 3-1 are based on the revenues reported by local companies owning more than 98 percent of all telephone access lines in the United States and an estimate of total interexchange revenues from companies reporting to the FCC. The other 1 percent to 2 percent of local telephone lines are owned by the small, usually rural carriers, for whom less precise data are available. Given that all facilities-based interexchange carriers are supposed to report their interexchange revenues to the FCC, the data on interexchange services should be reasonably complete.[2]

A decided slowdown has occurred in the growth of local service since 1982 (table 3-1). The output series for local services in 1984 reflects the detariffing of customer premises equipment (CPE) in that year. The growth in interstate MTS, however, has accelerated. Total growth in output of the telecommunications sector fell from about 8 percent a year in the 1970s to 3 percent between 1980 and 1985 but rebounded to more than 7 percent a year between 1986 and 1988.[3]

Unfortunately, no data are available on the output of nonregulated telecommunications services—principally private networks, smart buildings (buildings wired by the owner not by the telephone company) and specialized service carriers. A comparison of capital expenditures in the regulated sector with the estimate by the Bureau of Economic Analysis (BEA) of total capital expenditures in "telephone and telegraph" suggests that private systems and carriers have been substantial in the past ten years. According to the BEA data, the capital expenditures in telephone and telegraph tripled from almost $12.7 billion in 1975 to $38.6 billion in 1988 (table 3-2). However, reported capital expenditures by all local telephone companies, interexchange carriers, and telegraph companies rose only from $12.4 billion to $24.4 billion.[4]

The BEA data on capital spending are constructed by allocating

2. Many small interexchange carriers undoubtedly do not report their revenues to the Federal Communications Commission (FCC) either because they are too small to be required to report or because the FCC is simply unaware of their existence. These carriers without facilities are obviously only arbitragers who resell AT&T services that are captured in my data.

3. See chap. 5 for an analysis of this decline.

4. These data are compiled from FCC, *Statistics of Communications Common Carriers* (Washington, annual issues); United States Telephone Association (USTA), *Telephone Statistics* (Washington, annual issues); and annual reports of the new competitive long-distance carriers (OCCs).

Table 3-2. Alternative Measures of Telephone and Telegraph Plant and Equipment Investment
Millions of current dollars

Year	*Bureau of Economic Analysis (1)*	*Industry data (2)*	*(1) ÷ (2)*
1960	3,265	3,220	1.01
1961	3,356	3,293	1.02
1962	3,686	3,713	0.99
1963	4,051	3,876	1.05
1964	4,099	4,336	0.95
1965	4,812	4,917	0.98
1966	5,251	5,465	0.96
1967	5,348	5,708	0.94
1968	6,097	6,196	0.98
1969	7,408	7,427	1.00
1970	8,835	9,013	0.98
1971	9,063	9,909	0.91
1972	9,480	10,770	0.88
1973	11,913	12,028	0.99
1974	13,109	13,300	0.99
1975	12,683	12,379	1.02
1976	14,073	13,024	1.08
1977	17,747	14,942	1.19
1978	21,949	18,053	1.22
1979	26,800	21,180	1.27
1980	26,081	23,011	1.13
1981	29,217	24,402	1.20
1982	30,155	23,628	1.28
1983	31,960	22,312	1.43
1984	34,125	22,908	1.49
1985	36,674	25,366	1.45
1986	39,617	27,295	1.45
1987	38,481	26,052	1.48
1988	38,639	24,449	1.58

Sources: For 1960–85, Bureau of Economic Analysis (BEA), Fixed, Nonresidential Private Capital by Industry, Telephone and Telegraph, data base (June 1988). (Hereafter BEA private capital data base.) The 1986–88 data obtained from John Musgrave, BEA. The industry data are author's calculations based on FCC, *Statistics*, USTA, *Independent Telephone Statistics*, *Telephone Statistics*, and *Statistics of the Local Exchange Carriers, 1989*; and annual reports.

shares of the consumption of various categories of capital goods to the telephone and telegraph industry, SIC 481, 482, and 489, according to the 1977 Capital Flow Table, which reconciles plant and equipment spending with domestic capital-goods consumption.[5]

5. Information supplied by John Musgrave, Bureau of Economic Analysis (BEA).

Table 3-3. Capital Expenditures on Telecommunications Systems, 1980–88
Billions of current dollars

Year	*Total (Bureau of Economic Analysis)*	*Telephone and telegraph companies*	*Inside wiring charged directly to customers*	*Business and residential purchases of customer premises equipment*	*Business purchases of large private branch exchanges*	*Private networks*
1980	26.1	23.0	...	...	0.5	2.6
1981	29.2	24.4	1.0	0.1	0.6	3.1
1982	30.2	23.6	2.6	0.3	0.7	3.0
1983	32.0	22.3	3.2	2.2	1.2	3.1
1984	34.1	22.9	4.2	2.6	1.4	3.0
1985	36.7	25.4	4.8	3.0	1.5	2.0
1986	39.6	27.3	5.1	3.2	1.6	2.4
1987	38.5	26.0	5.1	3.5	1.8	2.1
1988	38.6	24.4	4.5	3.2	1.8	4.7

Sources: For 1980–85, BEA private capital data base; 1986–88 data obtained from John Musgrave, BEA. Author's calculations based on FCC, *Statistics*, USTA, *Telephone Statistics*, and *Statistics of the Local Exchange Carriers, 1989.*

Clearly, this method leads to a $13 billion overstatement in spending by the telecommunications *industry* in 1988, but the $13 billion does represent capital outlays on telecommunications by entities other than regulated telecom carriers.

The divergence between the BEA and industry data in table 3-2 may be attributed to four categories: privately held CPE; privately held, large private branch exchanges (LPBXs); investment in new inside wiring charged directly to the customer; investment by private networks in transmission, switching, and other equipment (table 3-3). Given that businesses and government now account for about 60 percent of telecommunications revenues, perhaps $1.9 billion of the $3.2 billion in private investment in CPE and $2.7 billion of the $4.5 billion of inside wiring may be assigned to government and business customers in 1988. The entire $1.8 billion of LPBX and $4.7 billion of private-network equipment are also attributable to business and government customers. Thus, total spending by government and business customers for private systems may be estimated at $11.1 billion in 1988, or about 29 percent of all spending on telecommunications capital. Another $3.1 billion, or 8 percent of total spending, may be assigned to residential investment in inside wiring and CPE.

Using BEA's estimates of the average life of equipment and structures and the reported share of each in total investment by regulated carriers, I have constructed real capital stock estimates from reported capital expenditures of all regulated local, interexchange, and telegraph carriers.[6] These data are reasonably consistent with BEA data until the mid-1970s, but thereafter they diverge from the BEA estimates markedly for the reasons just discussed (table 3-4). Had the industry data grown at the same rate as the BEA series between 1975 and 1988, the net capital stock in common carriers would have been approximately 33 percent greater in 1988.

These data suggest that the regulated common carriers are now accounting for a smaller and smaller share of U.S. telecommunications. There are no direct measures of the total value of telecommunications services emanating from private networks owned by airlines, banks, insurance companies, large manufacturers, or even large apartment buildings, but my estimate suggests that nearly 25 percent of all telecommunications net capital stock is now in the hands of someone other than a telephone company and that about 19 percent of it is in private business networks or communications systems. Given that private networks require far less right of way and other slowly depreciating plant to deliver services, the share of telecommunications output accounted for by these private systems may be far more than this 19 percent.

The Structure of Interexchange Markets

The 1982 modified final judgment decree established 161 local access and transport areas (LATAs) in which the divested Bell operating companies are permitted to offer long-distance service. The Bell operating companies are not allowed to offer service between LATAs. As a result, the U.S. interexchange market must now be divided between inter-LATA and intra-LATA services for the purpose of analysis. Inter-LATA service includes intrastate service between LATAs and interstate service. Interstate-inter-LATA markets are the most competitive and the least restricted because state regulators cannot impede entry into this market. The state regulatory commissions may limit competition in intrastate

6. This series is constructed by the perpetual inventory method using BEA's straight-line depreciation and retirement rates.

Table 3-4. Alternative Measures of Capital Stock in Telephone and Telegraph
Millions of 1982 dollars

	Bureau of Economic Analysis		*Industry data*	
Year	*Gross*	*Net*	*Gross*	*Net*
1960	107,407	67,717	107,408	67,717
1961	113,734	71,858	113,592	71,583
1962	120,775	76,583	120,728	76,261
1963	128,637	81,851	128,187	80,986
1964	136,481	86,886	136,602	86,489
1965	145,916	93,218	146,333	92,991
1966	155,958	99,870	156,963	100,088
1967	165,667	105,751	167,588	106,698
1968	176,140	112,086	178,401	113,173
1969	188,615	120,012	191,016	121,082
1970	202,912	129,359	205,724	130,726
1971	216,298	137,348	220,804	140,197
1972	228,961	144,446	235,726	149,233
1973	245,034	154,617	252,015	159,208
1974	261,304	164,616	268,658	169,154
1975	273,952	170,747	280,927	174,323
1976	286,504	176,760	292,126	178,418
1977	302,859	186,202	304,840	183,884
1978	323,484	199,601	320,391	192,279
1979	348,392	216,456	338,173	202,526
1980	370,342	229,728	356,184	212,892
1981	392,397	242,479	372,698	221,344
1982	411,930	252,278	385,557	226,012
1983	430,921	261,133	395,283	227,576
1984	450,160	270,015	389,905	222,869
1985	470,584	279,594	400,634	226,624
1986	491,865	289,960	411,869	234,899
1987	510,612	297,572	421,092	236,951
1988	525,317	301,337	437,780	226,879

Sources: For 1960–87, BEA, Fixed Private Capital by Legal Form of Organization and Industry, data base (June 1988); 1988 data obtained from John Musgrave, BEA. For industry data, author's calculations based on BEA methodology. FCC, *Statistics*, USTA, *Independent Telephone Statistics*, *Telephone Statistics*, and *Statistics of the Local Exchange Carriers 1989*; and annual reports.

service between LATAs in a multi-LATA state, but most now permit some facilities-based competition.[7]

The regulation of intra-LATA competition is much more complicated. Intra-LATA competition results in revenue losses for local-exchange companies and therefore reduces their ability to cross subsidize basic, monthly local access/exchange rates from supracompetitive toll rates. As a result, many states have been hesitant to allow it. Seven states expressly forbade intra-LATA, interexchange (facilities-based) competition, ten allowed only partial competition, six had decisions pending on the issue, and one had taken no action as of early 1990.[8] Thus, only twenty-six states allowed full intra-LATA competition six years after divestiture. However, forty-two states allowed the resale of WATS services (arbitrage) in early 1990.

Before divestiture, AT&T and the independent telephone operating companies controlled about 96 percent of the interexchange market.[9] MCI, Sprint, and the other common carriers (OCCs) had captured less than a combined 4 percent of the market (table 3-5). With divestiture, however, the OCC share has risen to 18.7 percent of all interexchange revenues and nearly 25 percent of the market for non-local-exchange carriers (largely inter-LATA).[10]

Partly because of state regulatory restrictions, the local exchange carriers, both Bell operating companies and independent telephone companies, control most of the intra-LATA market. In 1985 the intra-LATA market accounted for 22 percent of all toll revenues. The Bell operating companies accounted for three-quarters of this market, but the operating companies' share of the intra-LATA toll market may begin to shrink as more states allow intra-LATA resale and facilities-based competition.

The international market is among the more rapidly growing markets for U.S. interexchange carriers, but it still produces only about 5 percent of such revenues. AT&T still dominates this market because other companies have found it difficult to obtain connections

7. "Status of Intrastate Competition and Regulation (as of November 1, 1987)" *State Telephone Regulation Report*, Capital Publications, Alexandria, Va., November 1987, pp. 6–7.

8. "Intra-Lata Competition (East)," and "Intra-Lata Competition (West)," *State Telephone Regulation Report*, June 28, 1990, p. 3; and July 12, 1990, p. 3.

9. The independent operating companies still provide about $4 billion of intrastate long-distance service annually.

10. Recent preliminary data released by the FCC show that the OCC share of the market (exluding local carriers) rose to 33 percent in 1989.

Table 3-5. Market Shares of All Toll Revenues and of Non-Local-Exchange Carriers, by Company, 1978–88

Percent

Year	*Non-local-exchange carriers*				*Local-exchange carriers*	*Total*
	AT&T	*MCI*	*Sprint*	*Others*		
	Share of all toll revenues					
1978	99.4[a]	0.3	0.2	<0.1	...	100.0
1982	96.5[a]	1.8	0.9	<0.8	...	100.0
1985	67.2	4.2	2.8	3.7	22.1	100.0
1986	63.7	5.9	3.7	4.5	22.2	100.0
1987	60.8	6.8	4.5	4.8	23.1	100.0
1988	57.5	7.9	5.5	5.3	23.8	100.0
	Share of non-local-exchange carriers' revenues					
1985	86.3	5.5	3.5	4.7	...	100.0
1986	81.9	7.6	4.8	5.7	...	100.0
1987	79.1	8.8	5.8	6.3	...	100.0
1988	75.4	10.4	7.3	6.9	...	100.0

Sources: For 1978, MCI, Sprint, and other data are 1979 annual reports of miscellaneous common carriers form P reports to the FCC from MCI, Sprint, GTE Telenet, and United States Transmission Systems. For 1978 and 1982, AT&T data are from AT&T's 1978 and 1982 December monthly IB reports to the FCC; USTA, *Independent Telephone Statistics*, vol. 1, 1980, 1983, p.20, both issues; and FCC, *Statistics* 1978, 1982, table 1-6, both issues. For 1982, MCI, Sprint, and others data, and 1985–88 data, FCC, *Statistics*, 1988–89, table 1-3.

a. Includes all interexchange shares for local-exchange carriers before 1984.

to foreign postal, telephone, and telegraph administrations (PTTs). Nonetheless, the OCCs are now beginning to make serious inroads into even this market. Most of these foreign PTTs have been extremely hostile to new competitors, whom they feel are trying to undermine their reciprocal monopoly position with AT&T. Despite this hostility, MCI and Sprint have increased their international operations rapidly since 1987.

Local Service

The provision of local access and switching services has traditionally been the province of franchised local-exchange carriers (LECs). These services are often thought to be classic "natural monopoly" services, that is, one firm can produce the entire output less expensively than any two or more firms.

In recent years, however, private users have begun to develop their own switching and access nodes. Customer-owned private branch exchanges now number over 53,000. Each one is a local switch connected to the public switched network. Another 5,000 private networks also distribute traffic locally as well as over long distances. An estimated 128 cellular telephone systems were operating in 1986, 46 of them were owned by nontelephone companies.[11] By the end of 1988, cellular subscribers totaled more than 2 million, and the number of systems had quadrupled to 517.[12] Finally, several new fiber-optic networks have been built in major cities to offer voice-data services to large business customers.[13]

Despite the growth of switching outside the public network, the local-exchange companies still dominate the local access, switched-services markets. Each of the new technologies—private branch exchanges, cellular telephone, private microwave, or local fiber-optics networks—requires interconnection with the public switched network to reach more than a small share of local subscribers. Eventually, large fiber-optics networks may reach a fairly large share of business customers in concentrated urban areas, or the cellular telephone may spread from its current vehicle-related base to the general population. However, for the present, the local exchange carriers offer the only local switched service capable of connecting the geographically dispersed residential subscribers and smaller or medium-sized businesses.

The entry of new local competitors to serve large customers, including the interexchange carriers, has provoked concern over the uneconomic bypass of local exchange carriers because of inefficient pricing of regulated LEC services. In 1990 the principal telephone holding companies reported their estimates of total 1989 bypass for their franchise areas. The seven regional Bell operating companies plus GTE estimated that bypass via non-telco lines (facility bypass) reduced access revenues by $1.68 billion in 1989 while service bypass—the use of dedicated telco private lines rather than switched access—cost them another $2.07 billion.[14] The loss of revenues owing to the use of non-telco lines constituted about 8.7 percent of 1988 access revenues.[15] Thus, even by the companies' own estimates, only a small share of traffic is bypassing their circuits en route to subscribers.

11. Huber, *Geodesic Network*, table L-3.

12. Cellular Telecommunications Industry Association, U.S. Cellular Industry Survey, Washington, 1990.

13. These new local carriers are generally called "metropolitan area networks."

14. *Study and Report*, FCC CC Docket 78-72; and GTE and regional Bell operating company filings as reported in "Half of Reported Bypass Concentrated in a Handful of States," *State Telephone Regulation Report*, May 31, 1990, pp. 1–7.

15. Author's calculations, FCC, *Statistics*, 1988–89.

Table 3-6. Telephone Operating Company Access Lines and Revenues, 1977–88

	Bell operating companies		*Independent companies*	
Year	*Access lines (thousands)*	*Operating revenues (millions of dollars)*	*Access lines (thousands)*	*Operating revenues (millions of dollars)*
1977	75,125	34,595	18,345	7,115
1978	78,366	39,030	19,246	8,085
1979	81,400	43,238	20,078	9,240
1980	83,884	48,349	20,808	10,475
1981	85,987	55,283	21,429	12,206
1982	86,921	61,194	21,672	13,880
1983	89,042	64,027	22,331	14,875
1984	91,454	57,868	23,020	16,000
1985	93,945	61,180	24,330	17,245
1986	97,007	63,897	25,196	18,469
1987	100,244	64,981	26,481	19,920
1988	101,152	65,181	28,557	20,732

Sources: USTA, *Statistics of the Local Exchange Carriers, 1989,* pp. 2–3, and *Telephone Statistics,* 1986, pp. 2–3.

Although the threat of bypass by non-LEC circuits has become an important issue in the drive to reprice telecommunications services, apparently very little bypass of LEC facilities by the principal long-distance carriers occurred in the first few years after divestiture. Peter Huber estimates that non-LEC facilities were used for only 0.5 billion of 340.5 billion minutes of connections in originating or terminating interexchange calls during 1986.[16] This share may be increasing in recent years, but it is still not very large. Since 1984 the real output of interexchange carriers has grown at nearly the same rate as reported minutes of switched access on local telephone company lines.[17]

Industry data show a remarkable stability in the share of telephone lines owned by the Bell operating companies (table 3-6), but the independents' share of revenues has risen markedly in the past ten years.[18] The Bell companies suffered large revenue losses at divestiture when they were forced to detariff embedded CPE, but the independents have only recently detariffed their CPE. As I have

16. Huber, *Geodesic Network,* table IX-4.

17. FCC, Common Carrier Bureau, Industry Analysis Division, *AT&T's Share of the Long Distance Market: Third Quarter, 1989* (Washington, 1989).

18. Access lines are the local connections of voice-grade lines to residential and business subscribers.

shown, the growth in local services has slowed markedly in recent years, a worrisome slowdown to the divested Bell operating companies who are prevented by decree from moving into various other telephone-related businesses.

Approximately 900 subsidized rural companies with 5,700 exchanges and 5 million subscribers are among the country's local operating companies. These companies are afforded low-interest loans from the Rural Electrification Administration, the Rural Telephone Bank, and the Federal Financing Bank. Subsidized loans currently support nearly one-half of their assets. The share of assets supported by these loans is falling, however, as is the magnitude of the interest rate subsidy. In 1977 the average interest rate on the book value of the rural carriers' debt was only 54 percent of the rate paid by the carriers reporting to the FCC. By 1987 this subsidized rate had risen to 58 percent of the average rate paid by all carriers.[19]

Rural telephone companies are now coming under pressure from three changes in subsidy programs: the share of their assets subsidized by federal agencies is declining, the degree of interest rate subsidy is declining, and the revenues flowing from interexchange services to local companies in the form of access charges are declining.[20] For all of these reasons, one would expect rural carriers to become more cost conscious and perhaps to raise rates to their subscribers.[21] To deflect some of the political pressures from rural subscribers, however, the FCC has agreed to establish a "high-cost" fund from interexchange access charge revenues to continue the subsidies to the higher-cost companies.

Telephone Rates

During the 1960s and 1970s, the consumer price of telephone service rose at roughly one-half the rate of general inflation. This decline in the real cost of telephone service reflected the rapid rate

19. FCC, *Statistics* (1977, 1987), pp. 28–29, p. 1, respectively; and Rural Electrification Administration, *Statistical Report Rural Telephone Borrowers* (Department of Agriculture, 1977, 1987), pp. 9–11, tables 3, 5.

20. See the discussion on repricing in chap. 5.

21. See chap. 5 for an analysis of rural rates.

of technical change in the industry. Between 1981 and 1986, however, telephone rates rose more rapidly than the general price level. This reversal obviously fueled much of the political opposition to recent FCC policy initiatives that have followed the AT&T divestiture. Since 1986 telephone rates have resumed their historical tendency to rise much less rapidly than the general inflation rate.

Given the changes that have occurred in telephone service in recent decades, measuring the price of this service over time is exceedingly difficult. Touch-tone telephones have replaced instruments with rotary dials. The mix of interstate toll, intrastate toll, and local service has changed dramatically. Direct dialing has replaced operator-assisted long-distance calls. New features such as call forwarding and message waiting have appeared. Consumers now often own their telephone instruments rather than leasing them from their local operating company. Any price index for overall telephone service must grapple with all of these changes.

The only continuous indexes of national telephone rates are the Bureau of Labor Statistics' consumer price index (CPI) and producer price index (PPI). The CPI measures the cost of telephone service to consumers; the PPI measures the price of service offered by telephone companies.[22] The difference is important. Consumers pay excise taxes on telephone service and have recently begun to purchase their instruments and inside wiring separately and independently of the purchase of telephone service. As a result, the trends in the CPI and PPI for certain telephone service categories may differ.

Neither the CPI nor the PPI is adjusted for changes in the quality of service, and both suffer from the problem of changing product mix. The CPI for telephone service was not disaggregated into separate components for local and long-distance service until 1977. Moreover, the weights in the CPI are necessarily revised only periodically. The last two revisions occurred in 1977 and 1986. Between 1977 and 1986, the share of long-distance (interstate and intrastate toll) in total service increased and the number of consumers owning instruments rose. Although the Bureau of Labor Statistics has tried to adjust for consumer ownership of handsets,

22. For a detailed discussion of these price indices, see James L. Lande, *Primer and Source Book on Telephone Price Indexes and Rate Levels* (Washington: Federal Communications Commission, 1987).

probably both the change in service mix and in the share of consumers owning phones led to a slight overstatement in the increase in the CPI during the early 1980s.

Producer Prices

The PPI for telephone service is available only for individual types of service—interstate toll, intrastate toll, local residential service, and so on—not as an aggregate index of all telephone service. Moreover, until recently, the PPI for long-distance service had been based solely on AT&T rates. Since 1976 competitive interstate toll services have offered lower rates than AT&T, but this difference has narrowed quite a bit as the new competitors' market shares have increased. Thus, the rate of increase in the PPI for interstate long-distance service is probably biased somewhat upward between 1976 and 1986.

As the FCC began to admit entrants into private microwave services in the late 1950s and into private-line common carriage in the late 1960s and early 1970s, AT&T could have chosen to be accommodating, holding its interstate rates close to pre-entry levels. Any loss in business would force AT&T to respond by asking for rate increases elsewhere or even in the services subject to competitive assault. However, AT&T chose a different course, aggressively reducing real rates in the face of new competitive threats.

Data on interstate telephone rates before 1972 are rather sketchy, but the available information shows that average rates for message telephone service (MTS) fell by about 7 percent in nominal terms between 1965 and 1970 (table 3-7). In real terms (using the GNP deflator), MTS rates declined by 5.7 percent a year between 1965 and 1970, but the decline slowed greatly in the 1970s. Between 1970 and 1976, real interstate rates declined by 3.3 percent a year.

The entry of specialized common carriers into private-line service began in 1969. Even as late as 1975, however, these new carriers had failed to penetrate significantly the interstate private-line market. They accounted for only 3 percent of a $1.05 billion market in that year.

Part of the explanation for the slow growth in the market shares of specialized carriers may be found in AT&T's pricing of private-line services. Despite repeated rebuffs from the FCC, AT&T

Table 3-7. Nominal and Real Interstate Telephone Producer Price Indexes, 1960–89

1972 = 100

Year	*Switched MTS*		*WATS*		*Private lines*	
	Nominal	*Real*[a]	*Nominal*	*Real*[a]	*Nominal*	*Real*[a]
1960	107.7	162.1	n.a.	n.a.	n.a.	n.a.
1965	103.5	142.4	n.a.	n.a.	n.a.	n.a.
1970	96.7	107.1	n.a.	n.a.	n.a.	n.a.
1972	100.0	100.0	100.0	100.0	100.0	100.0
1973	102.9	96.7	101.7	95.5	100.1	94.0
1974	103.0	88.7	101.1	87.1	99.5	85.7
1975	111.7	87.6	102.5	80.4	103.4	81.1
1976	118.9	87.6	104.7	77.2	108.2	79.7
1977	120.6	83.3	105.1	72.6	108.4	74.9
1978	121.9	78.5	105.1	67.7	108.6	69.9
1979	120.8	71.5	105.1	62.2	108.5	64.2
1980	124.6	67.6	108.2	58.7	109.7	59.5
1981	137.5	68.0	116.3	57.5	133.5	66.0
1982	152.0	70.7	125.8	58.5	156.3	72.7
1983	153.4	68.7	127.1	56.9	157.0	70.3
1984	148.8	64.2	122.7	53.0	159.2	68.7
1985	143.3	60.1	115.8	48.6	165.9	69.6
1986	133.0	54.3	105.9	43.3	168.3	68.8
1987	111.9	44.3	97.5	38.6	168.3	66.6
1988	110.2	42.2	93.0	35.7	168.3	64.6
1989	108.3	39.9	87.9	32.4	168.3	62.0
	Average annual percent change					
1972–76	4.3	−3.3	1.1	−6.5	2.0	−5.7
1976–83	3.6	−3.5	2.8	−4.4	5.3	−1.8
1983–89	−5.8	−9.1	−6.1	−9.4	1.2	−2.1

Sources: For 1960–70, Christensen, Christensen, and Schoech, "Total Factor Productivity," p.13. For 1972–89, BLS producer price indexes reported in "Monitoring Report of the Federal-State Joint Board," in FCC CC Docket 87-339, December 1988, pp. 318–23; and January 1990, pp. 236–41.

n.a. Not available.

a. Deflated by implicit GNP deflator.

succeeded in holding nominal private-line rates virtually constant between 1972 and 1976 (table 3-7). In real terms, interstate private line rates fell twice as rapidly as interstate MTS. It appears, therefore, that AT&T's pricing recognized the new threat of entry in private lines, but that AT&T was less concerned about entry into switched services given the FCC's stance toward switched MTS. After 1976, however, AT&T's pricing strategy changed dramatically.

In 1974 MCI entered the forbidden world of switched MTS with its Execunet service. At first, AT&T could have reasonably expected the FCC to repulse this unauthorized entry into a regulated market, but by 1977 it was clear that entry into switched long-distance services would be permitted.

AT&T responded to the new entry by holding nominal interstate MTS rates virtually constant from 1976 to 1980. In real terms, this strategy resulted in a 6.5 percent annual decline in MTS rates—double the rate of decline between 1970 and 1976 (table 3-7). As in the case of private lines, AT&T had apparently moved rather vigorously to meet MCI's competitive challenge.

AT&T also responded to growing competition by reducing its WATS rates.[23] Between 1972 and 1988 real WATS rates fell by more than 6 percent a year (table 3-7). The increasing attractiveness of WATS compared with rates for MTS generated substantial growth in revenues from AT&T's larger customers who might have been contemplating building their own networks, using competitive private-line services, or even, at a later date, simply shifting to competitive interexchange carriers to obtain their switched services.

AT&T's responses to competitive entry in both private lines and switched services was extremely aggressive given the extent of the threat in the first few years. Even by 1982, the new carriers accounted for only slightly more than 5 percent of the total interexchange (switched and private-line) market despite an interim tariff that gave them local telephone connections at less than half the amount charged to AT&T calls.[24] Thus, at the dawn of divestiture, the new rules on competitive entry into interstate telephone services had had only a minor impact on AT&T's share of the market, but they probably affected interstate rates more dramatically.

Since 1980 real interstate MTS rates have continued to fall, largely in response to lower interstate access rates that are the result of shifting part of the cost of subscriber loops to local access rates through the subscriber-line charges.[25] Between 1980 and 1983, real interstate WATS rates flattened out after falling sharply over the previous eight years. But since 1983, real interstate WATS rates have fallen at an annual rate of 9.4 percent. Thus, competition, the

23. WATS is wide area telephone service, a discounted bulk offering of long-distance service offered to large customers.

24. This ENFIA (exchange network fee for interexchange access) tariff was negotiated between federal regulators, AT&T, and MCI. It was as much as 65 percent lower than the rates charged AT&T partly because MCI's connections were of lower quality.

25. See the discussion of regulatory changes in chap. 2.

repricing of access, and technological progress still provide the impetus for lower real interstate telephone rates.

The recent trend in intrastate rates is very different from the trend in interstate rates. After declining in real terms from 1972 through 1980, local rates began to rise rapidly. Since 1983 real residential local rates have risen by about 3 percent a year (table 3-8). Part of the explanation for rising local rates has been the FCC's imposition of the subscriber-line charge to defray part of the non-traffic-sensitive costs of subscriber connections, but the surge began between 1980 and 1983, before these subscriber-line charges were implemented.

Intrastate WATS and MTS rates have fallen much more slowly than interstate rates since 1983, undoubtedly because public utility commissions have attempted to keep intrastate long-distance rates high so as to limit the upward pressure on local rates caused by changes in federal regulatory policies described in chapter 2.

Consumer Rates

The real cost of telephone service to consumers, as measured by the CPI (deflated by the overall CPI index), has also fallen since the 1960s (table 3-9). Between 1964 and 1972, for example, the annual rate of decline in the real CPI for telephone service was 2.4 percent. Between 1972 and 1981, the year before the settlement of the AT&T case, this decline accelerated to an annual rate of 5.7 percent during a period of rising general inflation. Between 1981 and 1986, in a dramatic reversal, the real price index for telephone service rose at an annual rate of 2.8 percent, but this rise proved to be a temporary reversal. The real price of telephone service to consumers is once again falling at about 4 percent a year.

The turnaround in telephone rates evidenced in the CPI conceals the rather diverse trends in the various individual services. Between 1983 and 1987, real local rates (as measured by the deflated CPI index for local service) rose by 5.5 percent a year while real interstate toll rates fell by 10.8 percent a year and intrastate toll rates fell by 2.2 percent annually (table 3-9). The decline in interstate rates has been particularly dramatic since 1983, the last year before the divestiture. The rise in real local rates, however, began in 1981, not after divestiture. Obviously, these increases in local rates are

Table 3-8. Nominal and Real Intrastate Producer Price Indexes, 1972–89

1972 = 100

	Local service							
	Business		*Residential*		*Intrastate WATS*		*Intrastate MTS*	
Year	*Nominal*	*Real*[a]	*Nominal*	*Real*[a]	*Nominal*	*Real*[a]	*Nominal*	*Real*[a]
1972	100.0	100.0	100.0	100.0	100.0	100.0	100.0	100.0
1973	104.0	97.7	102.9	96.7	102.2	96.0	103.3	97.0
1974	111.1	95.7	108.8	93.7	108.4	93.3	107.7	92.7
1975	117.1	91.8	113.3	88.8	112.8	88.5	113.8	89.2
1976	123.9	91.3	118.9	87.6	123.4	90.9	125.6	92.6
1977	122.4	84.6	119.3	82.4	128.8	89.0	131.9	91.1
1978	126.0	81.1	122.1	78.6	133.6	86.0	132.0	85.0
1979	128.5	76.0	123.4	73.0	136.8	80.9	131.6	77.9
1980	132.9	72.1	128.0	69.5	139.6	75.7	132.3	71.8
1981	148.8	73.6	144.1	71.3	147.6	73.0	137.3	67.9
1982	162.7	75.7	160.6	74.7	149.9	69.7	145.6	67.7
1983	172.7	77.3	169.6	75.9	148.2	66.3	152.1	68.1
1984	200.4	86.5	182.4	78.8	147.8	63.8	157.0	67.8
1985	222.8	93.4	202.5	84.9	150.4	63.1	162.0	67.9
1986	232.9	95.2	223.6	91.4	151.8	62.0	158.0	64.6
1987	232.9	92.2	233.1	92.3	144.9	57.4	153.5	60.8
1988	230.7	88.4	236.6	90.7	135.0	51.8	149.3	57.2
1989	234.6	86.4	250.8	92.3	131.0	48.2	152.1	56.0
				Average annual percent change				
1972–83	5.0	−2.3	4.8	−2.5	3.6	−3.7	3.8	−3.5
1983–89	5.1	1.9	6.5	3.3	−2.1	−5.3	0.0	−3.3

Sources: BLS producer price indexes as reported in "Monitoring Report of the Federal-State Joint Board," December 1988, pp. 318–23, and January 1990, pp. 236–41.
a. Deflated by implicit GNP deflator.

Table 3-9. The Trend in Real Consumer Price Indexes for Telephone Services, 1964–89
1977 = 100

Year	*All telephone service*	*Local service*	*Interstate toll*	*Intrastate toll*
1964	152.2	n.a.	n.a.	n.a.
1965	147.4	n.a.	n.a.	n.a.
1966	140.4	n.a.	n.a.	n.a.
1967	138.2	n.a.	n.a.	n.a.
1968	132.7	n.a.	n.a.	n.a.
1969	127.5	n.a.	n.a.	n.a.
1970	121.8	n.a.	n.a.	n.a.
1971	122.5	n.a.	n.a.	n.a.
1972	125.2	n.a.	n.a.	n.a.
1973	121.0	n.a.	n.a.	n.a.
1974	113.6	n.a.	n.a.	n.a.
1975	107.4	n.a.	n.a.	n.a.
1976	105.2	n.a.	n.a.	n.a.
1977	100.0	100.0	100.0	100.0
1978	93.9	94.0	92.1	93.1
1979	84.2	84.0	82.2	84.5
1980	76.0	77.4	73.4	73.4
1981	75.0	78.2	72.1	68.5
1982	77.8	82.1	74.8	69.0
1983	80.3	86.1	74.0	71.5
1984	83.4	94.2	69.4	72.3
1985	83.7	98.0	64.1	70.4
1986	86.2	106.0	58.6	69.1
1987	82.6	107.4	48.2	65.4
1988	79.0	104.6	44.4	60.9
1989	76.1	103.5	41.0	55.5
		Average annual percent change		
1977–83	−3.7	−2.5	−5.0	−5.6
1983–89	−0.9	3.1	−9.8	−4.2

Sources: For 1964–84, Bureau of Labor Statistics (BLS). For 1985–89, BLS producer price indexes as reported in "Monitoring Report of the Federal-State Joint Board," December 1988, pp. 313–17, and January 1990, pp 234–35. All indexes are consumer price indexes deflated by the consumer price index, all urban consumers for all items.
n.a. Not available.

connected with other regulatory and economic forces and are not simply the result of divestiture.[26] Moreover, since 1986 real consumer rates for telephone service have fallen dramatically as even local rates have failed to keep pace with the overall CPI despite continuing increases in subscriber-line charges.

26. See chap. 5.

Who Pays for Telephone Services?

Most of the policy debate on telephone rates involves only the prices charged to consumers for local service, long-distance service, or equipment rental. The shifting of the burden of non-traffic-sensitive costs, for example, has aroused bitter opposition from consumer groups and organizations representing the elderly.[27] These debates rarely acknowledge that a very large share of telephone service is purchased by private business firms whose output prices reflect the cost of all of their inputs, including telephone service.

Since the mid-1970s the share of telephone revenues from common carriers accounted for by consumers has fallen from 49 percent to 42–44 percent (table 3-10). According to the data on telephone capital stock in table 3-4, nearly 25 percent of all telephone capital is in the hands of individuals or institutions other than common carriers. Thus, consumer purchases of telecommunications services from common carriers are now equal to about 34 percent of all services.[28] This figure suggests that two-thirds of all expenditures on telecommunications services are accounted for by business, by government, or by consumer outlays on telephone equipment.

The business and government spending for telecommunications services is heavily concentrated in service industries, such as retailing, wholesaling, finance, health services, and transportation. As table 3-11 shows, in 1977, roughly 60 percent of business spending on telephone service was concentrated in these service sectors. Given the widespread public use of these services and the competitive structure of these markets, any reductions in telephone costs are most likely quickly reflected in lower service prices. Of course, such effects are difficult to estimate directly given the small share of total costs accounted for by telephone services.[29]

27. See, for example, Mark N. Cooper and Mitchell Shapiro, "Low Income Households in the Post-Divestiture Era: A Study of Telephone Subscribership and Use in Michigan," Michigan Citizens' Lobby, October 1986; Gene Kimmelman and Mark Cooper, "Divestiture Plus Five: Residential Telephone Service Five Years after the Breakup of AT&T," Consumer Federation of America, Washington, December 1988.

28. If consumer expenditures are 43 percent of measured output and total output is at least 25 percent greater than measured output, consumer expenditures are actually only equal to 43 percent divided by 1.25 or 34 percent of total output.

29. The effect of repricing of telephone service on these sectors and, therefore, on consumers purchasing these services is examined in chap. 5.

Table 3-10. Consumer Spending on Telephone Services, 1965–88
Billions of current dollars

Year	*Personal consumption expenditures on telephone service (BEA)*	*Common-carrier telephone revenues*	*Consumer share*
1965	6.5	13.3	0.49
1966	7.0	14.7	0.48
1967	7.7	15.7	0.49
1968	8.3	17.2	0.48
1969	9.3	19.2	0.48
1970	10.1	20.9	0.48
1971	11.0	22.9	0.48
1972	12.4	25.9	0.48
1973	14.1	29.2	0.48
1974	15.5	32.5	0.48
1975	17.7	36.0	0.49
1976	19.8	40.7	0.49
1977	21.5	45.4	0.47
1978	23.9	51.3	0.46
1979	25.8	57.4	0.45
1980	27.8	64.4	0.43
1981	31.2	73.8	0.42
1982	35.6	82.5	0.43
1983	37.9	87.6	0.43
1984	39.8	89.4	0.45
1985	40.4	96.0	0.42
1986	42.7	101.7	0.42
1987	44.1	105.4	0.42
1988	47.2	107.4	0.44

Sources: BEA, Personal Consumption Expenditure Telephone Service, data base (October 1990). Author's calculations based on data from FCC, *Statistics*,; and USTA, *Independent Telephone Statistics, Telephone Statistics*, and *Statistics of the Local Exchange Carriers, 1989*. Adjustments were made for billing and leasing revenues to avoid double counting in the common-carrier telephone revenues, 1984–88.

Output, Employment, and Productivity in Common Carriers

Obviously, many entities other than common carriers now produce telephone services. Competition has made serious inroads into a once heavily regulated, monopolized industry. But how has this competition affected the output and productivity growth of common carriers? Has divestiture had marked effects on either output or

Table 3-11. Business Spending on Telephone Services by Input-Output Sector, 1977
Millions of current dollars

Sector	*Expenditure*
Wholesale and retail trade	5,584
Business services	2,619
Finance/insurance	2,525
Health, education, and social services	1,868
State and local government	1,636
Federal government	1,063
Transportation, warehousing	994
Telephone and telegraph	946
New construction	704
Hotels; personal and repair services	632
Printing and publishing	602
Real estate/rentals	510
Auto repair services	455
Eating and drinking places	449
Food and kindred products	365
Apparel	350
Maintenance and repair	348
Electric, gas, water, sewer service	326
Amusements	243
Other agriculture products	179
Petroleum refining	173
Miscellaneous manufacturing	161
Chemicals	153
Radio, TV, and communication equipment	147
Aircraft and parts	142
Livestock	139
Other fabricated metal products	128
Broad and narrow fabrics	126
Federal government enterprises	119
Screw machines	117
Heating and plumbing	115
Rubber and plastics	110
Drugs, cleaning products	109
General industrial machines	106
Total outlays for identified I-O sectors	23,297
Total business outlays	30,474

Source: *Survey of Current Business*, vol. 64 (May 1984), pp. 55–62.

productivity? To answer these questions, one must first construct series for output and inputs in the regulated telephone industry.

The total output of all regulated local exchange, interexchange, and telegraph companies is shown in table 3-12. This series is constructed by deflating output series for each type of service by the appropriate producer price index for 1972–88. For 1960–72 these

Table 3-12. Telephone Industry Indexes of Industry Output, 1960–88
1977 = 100

Year	*Industry data (1)*	*BLS gross output (2)*	*BEA gross product originating (3)*
1960	27.5	27.4	29.0
1961	29.1	29.1	30.4
1962	31.1	31.2	32.9
1963	33.3	33.5	35.9
1964	36.0	36.1	38.8
1965	39.1	39.4	42.3
1966	43.1	43.5	46.6
1967	46.4	47.2	50.6
1968	50.9	51.0	55.0
1969	56.2	56.6	60.5
1970	60.6	60.4	66.9
1971	64.2	63.4	70.2
1972	69.3	69.2	76.9
1973	76.1	76.2	83.1
1974	81.7	81.8	87.3
1975	85.6	84.8	90.7
1976	91.2	90.8	94.2
1977	100.0	100.0	100.0
1978	111.0	110.8	110.4
1979	123.2	122.2	118.4
1980	134.9	133.2	129.4
1981	142.5	140.7	136.7
1982	146.5	144.2	140.6
1983	149.9	146.6	151.5
1984	149.3	149.1	151.1
1985	156.6	153.7	156.1
1986	167.9	159.8	162.4
1987	183.8	167.9	177.5
1988	194.6	177.4	n.a.

Sources: For column 1, author's calculations of deflated revenues based on data from USTA, *Independent Telephone Statistics, Telephone Statistics*, and *Statistics of the Local Exchange Carriers, 1989*; and FCC, *Statistics*. For 1960–72, author's calculations include data from annual reports of other common carriers and Western Union; and Christensen, Christensen, and Schoech, "Total Factor Productivity," p. 13. For column 2, BLS, Telephone Communications, SIC 4811 Output, data base (July 1990). (Hereafter BLS SIC 4811 data base.) For column 3, BEA, Gross Product Originating, data base (September 1990).
n.a. Not available.

revenues are deflated by price indexes found in Laurits R. Christensen, Diane Cummings Christensen, and Philip E. Schoech.[30]

Whatever series one examines, clearly, the growth in the output of common carriers slowed substantially between 1980 and 1988.

30. See source for table 3-7.

Table 3-13 Employment and Labor Productivity for Telephone Communications, 1975–89

Year	Total employment (BLS) (thousands) (1)	Total employment (industry data) (thousands) (2)	Output per employee hour (1977 = 100) (3)	Value added per employee hour (4)
1975	966.6	960.2	85.9	89.2
1976	953.2	949.4	93.3	94.8
1977	957.3	962.7	100.0	100.0
1978	994.8	1,039.6	105.8	107.5
1979	1,048.2	1,067.8	110.8	115.2
1980	1,072.2	1,081.8	118.1	125.1
1981	1,077.3	1,092.5	124.4	130.7
1982	1,071.8	1,060.2	129.1	138.2
1983	956.0	950.9	145.1	158.9
1984	953.6	878.3	143.0	167.9
1985	920.9	852.6	149.8	175.4
1986	883.6	821.9	161.3	194.6
1987	902.3	811.4	165.9	218.7
1988	901.6	767.1	176.7	244.1
1989	876.9	n.a.	n.a.	n.a.

Sources: For column 1, BLS, Labstat Series Report: SIC 481 Telephone Communications, All Employees and Production Workers, data base (October 1990). For column 2, calculations are based on data from USTA, *Telephone Statistics*, FCC, *Statistics*, and AT&T's 1984–86 December monthly reports, no. 0001, to the FCC. AT&T's 1987–89 data obtained from Steve Friedlander, AT&T, and MCI annual reports. For column 3, BLS SIC 4811, Output per employee hour data base. For column 4, author's calculations based on data from FCC, *Statistics*, USTA, *Telephone Statistics*, and *Statistics of the Local Exchange Carriers, 1989*.
n.a. Not available.

In the same period, according to BLS estimates, employment in the telephone industry has declined by more than 16 percent (table 3-13). Much of this decline may be attributed to the deceleration in the industry's growth, but divestiture has surely added to the attrition of employment since 1981.

The BLS data on employment in the telephone industry match reported industry data quite closely until 1983. Thereafter, employment reported by the companies to the FCC and to the U.S. Telephone Association fall dramatically in comparison with BLS estimates.[31] Part of the reason for this divergence may be the shift of AT&T employees into unregulated subsidiaries whose data are not reported to the FCC. In addition, the diversified activities of

31. The data in column 2 in table 3-13 include all employment reported by independent companies to the USTA; AT&T and the regional Bell companies to the FCC; and the telegraph carriers to the FCC. In addition, they reflect the author's estimate of OCC employment.

local exchange companies may occur in establishments that are assigned to "telephone" by the Bureau of Labor Statistics. Whatever the reason, clearly, telephone companies have greatly reduced their employment in their traditional regulated activities. Between 1981 and 1988, the independent telephone companies reported a decline of 14 percent in employment; AT&T and the regional Bell companies reported a decline of 37 percent.[32] Very little of this decline of 490,000 people in regulated-company employment could possibly be offset by the new competitive carriers.

Labor productivity growth in the telephone sector has risen sharply since 1984 although the BLS statistics show a fairly stable pattern because they apparently include the employees of unregulated telephone-company subsidiaries. The BLS productivity series in table 3-13 grows at about 5.8 percent a year between 1975 and 1988. However, this study's estimates of industry value added and employment generate very different results, suggesting that value added for each worker hour grew at about 7 percent a year in the 1970s and 1980s.[33] A loglinear regression of employment on a time trend and industry output between 1960 and 1981 overpredicts 1988 employment by 19 percent.[34]

A better measure of productivity growth is the growth in total factor productivity (TFP) (table 3-14). To estimate the TFP for the entire telephone and telegraph industry, I construct estimates of real output, capital services, labor services, and materials use from data appearing in annual reports to the FCC and the United States Telephone Association as well as from annual corporate reports. The real output variable is the summation of real revenues shown in column 1 of table 3-12. The capital services are proportional to a weighted average of the net and capital stock of firms in the industry. These capital stock estimates are constructed from investment flows and BEA estimates of depreciation and retirement rates. Labor services are equal to estimated employees times average annual hours worked, adjusted for changes in the quality of the work force. Materials are a residual of revenues less labor costs less

32. USTA, *Statistics of the Local Exchange Carriers* (Washington, 1989), p. 3; and FCC, *Statistics* (1981, 1989), pp. 32, 39, respectively.

33. Value added is total industry real output minus real material purchases.

34. The regression is LogEmployment = $a_0 + a_1$ LogOutput + a_2 time, where Employment is the industry estimate from column 2 in table 3-13 and Output is the estimate from column 1 in table 3-12.

Table 3-14. Annual Growth in Output, Labor Input, Capital Input, and Total Factor Productivity in Regulated Telephone Common Carriage, 1961–88
Percent

			Labor hours				*Total factor productivity*	
Year	*Output*[a]	*Value added*[b]	*Unweighted*[c]	*Weighted*[d]	*Capital stock*[e]	*Materials*[f]	*3-factor*[g]	*2-factor*[g]
1961	5.8	5.5	−3.7	−2.8	5.7	6.6	3.0	3.6
1962	6.8	6.5	0.9	1.2	6.3	6.0	2.3	2.4
1963	7.2	7.4	1.5	2.0	6.2	4.9	2.7	3.0
1964	7.8	7.7	3.5	3.4	6.6	6.6	2.2	2.5
1965	8.7	8.3	4.2	3.6	7.2	6.1	2.9	2.6
1966	10.3	10.2	7.0	6.2	7.3	6.7	3.5	3.3
1967	7.7	7.7	−2.0	−2.1	6.7	4.5	4.4	4.7
1968	9.6	9.4	5.0	4.8	6.4	4.3	4.1	3.7
1969	10.5	7.7	9.4	8.3	7.1	16.1	1.1	0.1
1970	7.8	8.7	2.6	2.4	7.7	0.2	3.6	3.2
1971	5.9	5.4	−2.8	−1.4	7.3	5.1	2.1	2.1
1972	7.9	7.7	3.8	5.9	6.7	9.1	1.0	1.3
1973	9.8	11.0	3.3	3.2	6.9	2.0	5.3	5.8
1974	7.4	7.1	−0.8	−0.1	6.5	3.2	4.0	3.6
1975	4.7	5.3	−4.9	−2.8	4.3	−1.1	4.1	4.2
1976	6.5	4.8	−1.1	0.3	3.7	12.8	2.0	2.6
1977	9.7	7.8	3.7	5.9	4.1	10.8	3.4	2.8
1978	11.0	13.4	6.1	6.2	5.0	−1.8	7.1	7.8
1979	11.0	11.0	5.3	4.6	5.5	2.4	6.5	6.1
1980	9.5	8.2	1.3	0.1	5.3	6.8	5.7	5.4
1981	5.6	4.8	0.7	1.0	4.5	7.8	1.6	1.9
1982	2.8	1.8	−3.7	−3.3	3.2	7.9	0.8	1.8
1983	2.3	3.8	−10.1	−9.4	2.2	−3.1	5.5	7.3
1984	−0.4	−2.4	−6.5	−6.5	−1.5	3.7	1.6	1.9
1985	4.8	2.3	−1.2	−1.1	2.6	13.6	0.6	0.6
1986	7.2	5.1	−2.9	−2.7	2.9	4.8	5.4	5.9
1987	9.5	8.9	−1.3	−1.5	2.0	2.9	8.2	7.8
1988	5.8	4.2	−5.7	−6.6	2.7	9.0	3.8	2.9
				Average annual growth				
1961–70	8.2	7.9	2.8	2.7	6.7	6.2	3.0	2.9
1971–83	7.2	7.1	0.1	0.8	5.0	4.8	3.8	4.1
1984–88	5.4	3.6	−3.5	−3.7	1.7	6.8	3.9	3.8

a. Deflated real revenues. Revenues from FCC, *Statistics*, USTA, *Independent Telephone Statistics*, and *Telephone Statistics*, and *Statistics of the Local Exchange Carriers, 1989*. For 1972–88, deflators are BLS producer price indexes; for 1961–71, AT&T deflators are from Christensen, Christensen, and Schoech, "Total Factor Productivity."

b. Deflated real revenues less real material costs.

c. Industry employment times average hours worked. Employment data from FCC and USTA statistics and company annual reports. Hours from BLS, *Employment and Earnings*, yearly March issues, for SIC 481.

d. For 1960–79, total hours adjusted for labor quality per Christensen, Christensen, and Schoech, "Total Factor Productivity"; for 1980–88, estimates are based on predictions from a regression of the labor quality measure on real industry wages in 1960–79.

e. Constructed from industry investment flows obtained from FCC and USTA statistics and company annual reports. Depreciation and retirement rates from John Musgrave, BEA. Capital stock is a weighted average of net and gross capital, using 0.25 and 0.75, respectively, as weights. See Edward Denison, *Accounting for Slower Economic Growth: The United States in the 1970s* (Brookings, 1979), pp. 50–52.

f. Revenue less capital costs and labor costs from FCC and USTA statistics and company annual reports, deflated by GNP implicit price deflator.

g. Tornquist-weighted TFP, using cost shares as weights from FCC and USTA statistics and company reports.

(accounting) capital charges. Value added is simply real output less real material costs. TFP estimates are constructed using both gross output and value added as output measures, resulting in three-factor and two-factor measures of TFP, respectively.[35] The discrete rates of change in each variable are shown in table 3-14.

The resulting Tornquist-weighted estimates of TFP grow much more rapidly between 1971 and 1983 than between 1961 and 1970, the period of no competition in interexchange services. Estimated declines in total factor productivity 1984–85, the year of divestiture, could simply reflect measurement problems or transitory inefficiencies. The divestiture changed the structure of the Bell operating companies and AT&T so radically that possibly there are inconsistencies in data reported to the FCC before and after 1984.[36]

After fairly modest increases between 1984 and 1985, total factor productivity accelerated between 1986 and 1988. Thus, TFP now seems to be growing more rapidly than in the 1970s. In fact, TFP growth has accelerated over the entire postliberalization period of 1971–88.

Because telephone companies must fulfill a public-service responsibility, they are required to maintain a certain excess capacity and are generally forbidden to ration excess demand through rate

35. The value-added measure is theoretically inferior to a gross output measure if materials are not weakly separable from capital and labor in the production function. See Ferenc Kiss, "Productivity Gains in Bell Canada," in Leon Courville, Alain de Fontenay, and Rodney Dobell, eds., *Economic Analysis of Telecommunications* (Amsterdam: North Holland Press, 1983), p. 97.

36. It is reassuring, however, that the average TFP growth between 1961 and 1980 reported in table 3-12 is 3.5 percent. This figure compares with an average of 3.44 percent reported by Kiss, "Productivity Gains," for Bell Canada (p. 96) and 3.3 percent obtained by Laurits R. Christensen, Diane Cummings Christensen, and Philip E. Schoech, "Total Factor Productivity in the Bell System, 1947–79," Christensen Associates, Madison, Wis., September 1981 for AT&T between 1961 and 1979. The methodology employed in this study departs from Christensen, Christensen, and Schoech in one important respect. Christensen constructs a Divisia index of output, using revenue shares as weights. Given the distorted rates for telephone service, this approach is not very satisfactory. In this study, the output measure is equal to the sum of deflated revenues, which is at best a rough approximation of real output. In research not reported here, I have attempted to construct a Divisia index using adjusted revenue-share weights that account for regulatory distortions. The resulting estimates of TFP growth are slightly lower—they average 3.05 percent over the 1961–80 period—but the temporal pattern of the results is the same. See Robert W. Crandall and Jonathan Galst, "Productivity Growth in the U.S. Telecommunications Sector: The Impact of the AT&T Divestiture," November 1990.

increases. This means that measured TFP rises and falls with cyclical swings in output, but input levels are not at all times adjusted to these output levels. Therefore, one should expect to see TFP growth rise and fall with changes in output growth.

The data in table 3-14 suggest that TFP growth has generally trended upward over the 1961–88 period of the analysis, but that it dipped in 1984–85 because of divestiture. A simple regression over this 1961–88 period of three-factor TFP growth ($\dot{TFP}/TFP$) on output growth ($\dot{Q}/Q$), a dummy variable for 1984–85 (D84-85), and a time trend (TIME) yields the following result:

(1)

$$\dot{TFP}/TFP = -0.606 + \underset{(2.51)}{0.327}\,\dot{Q}/Q - \underset{(1.49)}{2.09}\,D84\text{–}85 + \underset{(3.15)}{0.123}\,TIME$$

where the numbers in the brackets are the absolute value of t-statistics.[37] This result suggests that productivity growth accelerated at a rate of 0.12 percent a year between 1961 and 1987, but deviated from this trend by 2.1 percent a year from 1984 to 1985, the divestiture period. However, when the TIME trend is replaced by a TIME(72) variable that assumes a value of unity between 1961 and 1971 and then 2, 3, 4 . . . 18 from 1972 to 1988 to capture the temporal trend after the 1971 Specialized Carrier decision, the results improve dramatically.[38]

(2)

$$\dot{TFP}/TFP = -0.021 + \underset{(3.00)}{0.367}\,\dot{Q}/Q - \underset{(1.63)}{2.15}\,D84\text{–}85 + \underset{(3.77)}{0.19}\,TIME(72)$$

The addition of a trend for 1961–71 to (2) does not add to the explanatory power of the regression, suggesting no upward trend in TFP growth in this preliberalization period. Equation (2) allows one to conclude that by 1988, TFP growth is fully 3.2 percentage points above its 2.9 percent average for 1961–71 after cyclical effects are eliminated. Thus, one is left with the conclusion that something occurred just after 1970 that began to elevate productivity growth in this industry. It may have been a series of technological advances or, given the technological breakthroughs that also occurred in the 1950s and 1960s that failed to accelerate TFP growth between 1961

37. $\bar{R}^2 = 0.336$, $\rho = -0.062$, and Durbin Watson = 2.011.
38. $\bar{R}^2 = 0.388$, $\rho = 0.133$, and Durbin Watson = 2.110.

Table 3-15. Average Hourly Earnings for Production or Nonsupervisory Workers, 1972–89
Dollars per hour

Year	*Total private (nonfarm)*	*Telephone communications services*	*Telephone and telegraph equipment*
1972	3.70	4.10	4.18
1973	3.94	4.45	4.45
1974	4.24	4.90	4.73
1975	4.53	5.61	5.23
1976	4.86	6.33	5.78
1977	5.25	6.89	6.28
1978	5.69	7.50	6.73
1979	6.16	8.07	7.50
1980	6.66	8.72	8.24
1981	7.25	9.80	9.15
1982	7.68	10.65	9.86
1983	8.02	11.41	10.27
1984	8.32	11.91	10.69
1985	8.57	12.45	10.82
1986	8.76	12.86	10.94
1987	8.98	13.19	11.17
1988	9.29	13.63	11.62
1989	9.66	13.88	11.68
	Average annual percent change		
1972–81	7.5	9.7	8.7
1981–89	3.6	4.4	3.1
1983–89	3.1	3.3	2.1

Sources: For 1972–78, BLS, *Employment and Earnings, United States, 1909–1978* (Department of Labor, 1979); for 1979–87, BLS, *Supplement to Employment and Earnings Revised Establishment Data* (Department of Labor, annual issues); for 1988–89, BLS, Total private and telephone and telegraph equipment, BLS, *Employment and Earnings* (Department of Labor, annual issues), table C-2; and for 1988–89 telephone communications services, BLS, Labstat Series Report: SIC 481 Telephone Communications, Average Hourly Earnings, data base (October 1990).

and 1971, it is likely to reflect the boost to efficiency spurred by competitive entry. Moreover, the sharp decline in employment that began in 1982 has propelled TFP growth between 1982 and 1988, particularly after 1985. This suggests that divestiture stimulated a big improvement i efficiency as the Bell companies streamlined their operations.

Wages

Earnings for nonsupervisory workers in telephone communications services (table 3-15) grew by an average of 9.7 percent a year from 1972 to 1981, far outpacing the average growth in private, nonfarm wages. Since 1981, however, the rate of increase has been slowing, but it remains above the average for the private economy. From these data, it would appear that liberalization and divestiture have not yet seriously affected wage rates in the regulated telephone industry. In the equipment industry, however, wage rate increases have begun to slow markedly in comparison with the rest of the economy. This difference is undoubtedly because of the sharp rise in competition in equipment markets while local exchange markets continue to be sheltered from competition by state regulators.[39]

Summary

The introduction of competition and changes in federal regulatory practices have combined to create important changes in the relative prices of telephone services in the 1970s and 1980s. Local rates have risen in real terms since 1981, but interexchange rates have fallen dramatically in response to the new competition and regulatory repricing. Not surprisingly, the rate of growth of long-distance services has far exceeded the growth in local service.

The development of private networks and the rapid growth in business use of long-distance service have dramatically transformed the telephone industry. Today, only about one-third of all telecommunications services is accounted for by telephone company services provided for personal consumption. Most of the other two-thirds of telecommunications output is consumed by businesses and government.

As a result of competition and divestiture, the growth in the output of common-carrier telecommunications companies, particularly the local telephone companies, has slowed dramatically. The capital stock in these carriers is growing very slowly, and employment in these companies has fallen by nearly 30 percent since 1981. These changes do not mean that the nation's telecommunications system is declining, but that an increasing share of the service is now being delivered by unregulated private networks and equipment. As with transportation regulation in earlier decades, the combination of regulation and the freedom to purchase one's own equipment has resulted in the rapid growth in private carriage.

39. See chap. 4.

Much of the stagnation in regulated-carrier output is concentrated in the local telephone carriers. Under the AT&T decree, the divested Bell operating companies may not offer long-distance service outside certain court-drawn boundaries, manufacture equipment, nor offer most information services. As long as these restrictions remain, the Bell operating companies will not be able to grow, nor will they be able to offer many of the services permitted by the rapidly changing technology.

Through 1988 it appears that competition, divestiture, and the growth in private carriage did not reduce the efficiency of regulated carriers. Labor productivity continues to grow rapidly despite the slowdown in regulated-company output. Total factor productivity slowed noticeably in 1984 and 1985 but reaccelerated between 1985 and 1988. Since 1971 TFP growth has increased by 0.2 percentage points a year after failing to increase in the 1960s. This increase suggests that liberalization has improved the efficiency of the telecommunications system by more than any potential loss in scale or scope economies.

Chapter Four

The Equipment Market

THE REGULATORY and antitrust upheavals in the telephone industry have revolutionized the U.S. telephone equipment market. The dominant suppliers in the 1960s are now under substantial competitive pressure. The Bell operating companies are no longer required to acquire their equipment from AT&T's manufacturing subsidiary. Residential and business customers are buying terminal equipment from scores of different manufacturers. Equally important, the rate of introduction of new types of equipment, such as facsimile machines or portable cellular transceivers, has accelerated markedly.

Competitive Entry and the Equipment Markets

The introduction of entry into long-distance communications began with the Above-890 decision in 1959. In this decision, the Federal Communications Commission (FCC) opened the door for private transmission facilities and even private networks.[1] However, these private systems did not flourish in the 1960s, probably because of AT&T's aggressive pricing of Telpak (AT&T's discounted private-line service for large customers) and other private-line services.

When the FCC admitted specialized carriers in 1969 and 1971 it ushered in a new set of national carriers who were usually not affiliated with large suppliers of telephone equipment. These new

1. The commission opened this door only a crack in 1948 by allowing limited construction of microwave facilities by television broadcasters who were not served by AT&T circuits.

carriers built their networks at first with microwave transmission facilities and offered digital data services. Subsequently, they invested heavily in large fiber-optics networks. As a result, they generated a demand for rather different equipment from that typically used by AT&T or the independent telephone companies before 1980.

In the 1970s firms began to offer various value-added services, such as packet-switching, the switching and transmission of packets of information (principally data) so as to economize on transmission capacity. These new networks might use common-carrier transmission facilities, but they would invest in specialized equipment to gather, transform, and send signals to the common-carriage network. Financial institutions, airlines, retailers, and other large private corporations began to build their own highly specialized networks in the 1970s and 1980s, networks that would have violated FCC regulations before 1959.

The development of fiber-optics technology is creating an entirely new wave of entry into communications carriage. Large national firms and consortia are building very large regional and national networks to compete in the interexchange market. In addition, numerous local networks are being built with fiber optics to compete with (or bypass) local telephone companies for the local access and transport business of large users, especially in major metropolitan areas. These new networks use equipment that is very different from the embedded plant of most existing telephone companies.

The other major new entrants into the local access/transport market are the cellular (mobile) telephone companies first authorized by the Federal Communications Commission (FCC) in 1982. These companies offer mobile communications that are connected to the common- carrier network, using radio transmission equipment and solid-state transceivers.

Finally, the advent of domestic satellite communications in the 1960s has led to a proliferation of private systems using satellites. Much of this market is devoted to the distribution of broadband (television) signals, but satellite systems are also used for a variety of voice/data services.

It would not be an exaggeration to say that before 1959, the U.S. telephone equipment market was dominated by Western Electric, selling to Bell operating companies and AT&T's Long Lines, and a few other small suppliers usually affiliated with the independent telephone companies.[2] No one would enter the telephone equipment market under these conditions because the established telephone companies were virtually the only equipment buyers.

Between 1959 and 1971 federal regulatory policy changed this situation dramatically, permitting new entry into services and opportunities for new equipment suppliers to sell to the new entrants. Liberalization of markets for terminal equipment in the mid-1970s opened the floodgates to entry of a myriad of new domestic and foreign suppliers of customer premises equipment (CPE). Finally, the 1984 AT&T divestiture that divorced Western Electric from the Bell operating companies created a much more competitive market for switching, transmission, and terminal equipment, as the Bell operating companies were relieved of the necessity of buying from Western Electric. In just twenty-five years, the structure of the equipment-buying industry, the technology of transmission and switching, the degree of vertical integration, and the nature of the services offered by this equipment have been thoroughly transformed by changes in regulatory policy that have stressed entry and competition rather than market stability and the preservation of natural monopoly.

Changes in U.S. Communications Equipment

To see how much the market for telecommunications has changed in the past ten years, one has only to examine the divergent trends in telephone-equipment consumption and telephone-company investment (table 4-1). Between 1977 and 1988 the consumption of telephone and telegraph equipment (SIC 3661) and nonbroadcasting communications systems and equipment (SIC 36631) rose from $10.4 billion to $27.2 billion, a 162 percent increase, while the total capital expenditures of all local telephone companies, long-distance carriers, and telegraph carriers rose by only 64 percent. Clearly, much of the increase in spending on communications equipment has come from entities other than traditional telephone companies.

One might think that much of the divergence since 1975 in the series in table 4-1 is because of the growth in consumer ownership

2. Western Electric became AT&T Technologies after the 1984 divestiture.

Table 4-1. Capital Expenditures on Telephone and Telegraph Systems, 1977–88

Billions of current dollars

Year	*Apparent consumption of telephone and telegraph equipment SIC 3661* (1)	*Apparent consumption of nonbroadcast communications equipment SIC 36631 (formerly 36621)* (2)	*Total capital expenditures by telephone carriers* (3)	*BEA estimate of telephone and telegraph industry capital expenditures* (4)
1977	6.93	3.47	14.94	17.75
1978	7.99	3.69	18.05	21.95
1979	9.59	4.03	21.18	26.80
1980	11.02	4.85	23.01	26.08
1981	11.76	5.50	24.40	29.22
1982	11.24	6.24	23.63	30.16
1983	11.98	7.20	22.31	31.96
1984	15.48	9.20	22.91	34.12
1985	17.22	10.66	25.37	36.67
1986	17.03	11.19	27.29	39.62
1987	18.18	11.69	26.05	38.48
1988	14.52[a]	12.65	24.45	38.64
		Percent growth		
1977–88:	110	265	64	118

Sources: For columns 1 and 2, Bureau of the Census, "Selected Electronic and Associated Products, including Telephone and Telegraph Apparatus, 1978, 1979, 1980, 1981, 1982, 1983, 1984," *Current Industrial Reports*, series MA36N (Department of Commerce, 1980–85), and "Communications Equipment including Telephone, Telegraph, and Other Electronic Systems and Equipment, 1985, 1986, 1987, 1988," series MA36P (Department of Commerce, 1986–90). For column 3, author's calculations based on data from Federal Communications Commission (FCC), *Statistics of Communications Common Carriers* (Washington, annual issues); and United States Telephone Association (USTA), *Telephone Statistics*, vol. 1 (Washington, annual issues), *Statistics of the Local Exchange Carriers, 1989*; and annual reports of MCI, U.S. Sprint, United Telecom, Alltel, Western Union, and AT&T to the FCC. For column 4, 1977–85, Bureau of Economic Analysis (BEA), Fixed, Nonresidential Private Capital by Industry, Telephone, and Telegraph, data base, (June 1988) and 1986–88 data obtained from John Musgrave, BEA.

a. Excludes modem imports.

of CPE, but as tables 4-2 and 4-3 demonstrate, there must be another explanation. Table 4-2 provides a breakdown of the major categories of telephone and nonbroadcast communications equipment in three discrete years: 1988, 1982, and 1978. The final consumption of telephone sets (instruments) has risen at about the same rate as switching equipment—a category that includes central office equipment and private branch exchanges. And the consumption of all other telephone and telegraph apparatus has grown at one-fourth the rate of telephone sets. However, since 1978, apparent consumption of nonbroadcast communications systems and equip-

Table 4-2. U.S. Manufacturers' Shipments, Exports, and Imports of Telephone, Telegraph, and Communications Equipment, Selected Years, 1978–88

Millions of current dollars

Category	*Shipments*	*Exports*	*Imports*	*Apparent consumption*
		1988		
SIC 36611-2				
Switching equipment (including PBX)	7,067	530	461	6,998
Telephone sets	359	56	1,400	1,634
Other telephone and telegraph equipment	3,173	562	2,125	4,736[a]
SIC 36631				
Mobile equipment (except amateur and citizens band)	1,944	248	278	1,974
Other communications systems and equipment except broadcasting[b]	10,975	1,468	1,172	10,678
		1982		
SIC 36611-2				
Switching equipment (including PBX)	4,448	371	184	4,261
Telephone sets	1,065	24	149	1,190
Other telephone and telegraph equipment	5,915	417	295	5,793
SIC 36631: (formerly 36621)				
Communications systems and equipment except broadcasting	6,854	947	336	6,243
		1978		
SIC 36611-2				
Switching equipment (including PBX)	3,231	93	88	3,226
Telephone sets	824	10	42	856
Other telephone and telegraph equipment	4,072	278	118	3,912
SIC 36631 (formerly 36621)				
Communications systems and equipment except broadcasting	3,810	704	580	3,685

Sources: Bureau of the Census, "Selected Electronic and Associated Products, 1979, 1983," series MA36N, and "Communications Equipment, 1988," series MA36P.

a. Excludes modems.

b. Excludes citizens band equipment.

Table 4-3. Shipments by U.S. Manufacturers of Telephone, Telegraph, and Communications Equipment, Selected Years, 1974–88
Millions of current dollars

Equipment	*1974*	*1978*	*1982*	*1988*
Telephone and telegraph equipment				
Central office switching equipment	1,700	1,694	1,959	4,612
PBXs	298	531	1,048	1,009
Carrier line equipment	[a]	773	1,563	4,512
Telephone sets	351	824	1,065	359
VF equipment	201	257	213	139
Display terminals	[a]	155	72	[a]
Teleprinters	[a]	276	334	[a]
All other telephone and telegraph equipment	2,291	3,617	5,174	5,550
Nonbroadcast communications systems and equipment				
Fiber-optics equipment	37 (Fiber-optics and other light transmission equipment combined)	45 (Fiber-optics and other light transmission equipment combined)	88	432
Other light transmission equipment			107	432
Transmitters, receivers point-to-point radio equipment	152	649	1,290	1,994
Telemetering equipment	86	115	192	319
Satellite equipment	289	396	1,614	2,217
Mobile systems and equipment	497	760	860	1,925
Portable receivers and other equipment	155	375	510	1,205

Sources: Bureau of the Census, "Selected Electronic and Associated Products, 1975, 1979, 1983," series MA36N, and "Communications Equipment, 1988" series MA36P.
a. Included in "all other telephone and telegraph equipment."

ment has risen more rapidly than consumption of traditional telephone equipment. The domestic suppliers account for about 90 percent of domestic consumption of these nonbroadcast communications systems, thus allowing one to examine U.S. shipments data, for which there is more detail, to draw further inferences about the directions of the communications equipment market.

The explanation for the rise in the nonbroadcast communications equipment sales is the sharp increase in fiber-optics systems, point-to-point radio systems (principally microwave), mobile systems, and satellite equipment (table 4-3). These are not the types of equipment bought by residential consumers or even small businesses (except for the smaller satellite dishes and some of the mobile equipment). Rather, the rapid growth in the sale of this equipment indicates a growing importance of newer technologies in building private and common-carriage systems. Even carrier-line equipment seems to be growing more rapidly than other important categories of telephone equipment. The surge in shipments of central office switching equipment since 1982 is partly because of the Bell operating companies' response to the requirement in the 1982 modified final judgment that they convert to equal access and to a renewed effort by the Bell operating companies to develop Centrex service as an alternative to private branch exchanges (PBXs).

The modest growth in U.S. shipments of most of the other categories of traditional telephone equipment suggests that new technologies and new private and public networks are supplanting traditional telephone common carriage. Even PBX sales seem to be growing very slowly since the late 1970s as telephone common carriers blunt the private demand for these decentralized switches through new Centrex tariffs and businesses develop their own local area networks for data communications.[3]

The Western Electric-AT&T Relationship

Western Electric retained its virtual monopoly of central office equipment until 1980. The Bell System accounted for more than 80 percent of the purchases of central office equipment, and Western Electric supplied most of the Bell System needs. The decline of AT&T in supplying the central office market in the early 1980s may be attributed more to advancing digital technology than to regulatory efforts.

AT&T consistently pursued the policy of having Western Electric supply most of its major central office equipment. The 1956 consent decree allowed the continued use of the Standard Supply Contracts between the Bell operating companies and Western Electric. Under these contracts Western Electric provided the operating companies with most of their purchasing and warehousing. In addition, these contracts gave Western access to operating companies' construction forecasts and network design information.

This relationship meant that if a Bell operating company wanted

3. Centrex is a multiline service that provides a multistation business subscriber with direct connections for each line to the central office switch, a single directory number for the subscriber, and identification of the outgoing station making the call. Centrex is thus a substitute for customer premise switching through a private branch exchange (PBX).

to purchase a non-Western switch, AT&T-Western Electric would know about it because it would be purchased through Western. It was alleged in the 1974 *U.S.* v. *AT&T* antitrust suit that this knowledge about future central office needs allowed Western substantial lead time to design equipment that could compete with the outside equipment. If AT&T could not convince a Bell company to wait for Western equipment, its inspection of the purchased equipment could at least help it design a competing switch to prevent future sales. Even if Western lost a sale, it acted as purchasing agent for the Bell operating companies and would buy the switch and charge the company a substantial markup.

For most of the period from 1956 to divestiture the Bell purchases of central office equipment manufactured by others were usually limited to fairly small switches and auxiliary equipment. The introduction of digital switches in the late 1970s sparked the first big change in this pattern of Bell company purchases. In 1976 and 1977, Northern Telecom, a Canadian company, introduced the first digital, time-division local office switches, the DMS-1 and the DMS-10. The former was a remote switch and the latter a small switch principally for rural use. In 1978 and 1979 Northern Telecom rounded out its line of digital, time- division switches with the DMS-200, a toll switch, and DMS-100, a local switch. The Bell operating companies took a great interest in these switches because AT&T had no equivalent offerings in the local-office switch market for local offices.[4] In 1980 Northern Telecom began to sell the DMS-10 and DMS-100 to the Bell operating companies, and it was not until 1982 that Western Electric was ready with a competitive offering, the #5ESS, which apparently had some technical problems when first introduced. The decision by the Bell operating companies to defect to Northern Telecom central office equipment was driven by customer demands and the need to thwart their customers' potential shifts to more sophisticated terminal equipment.

The growth of competition in the market for central office equipment was also abetted by the development of the new independent interexchange carriers in the late 1970s and early 1980s. These companies, although relatively small, needed advanced networks, and the non-Western manufacturers were ready to meet their needs.

4. Telephone conversation with Derek Davies, Northern Telecom, March 22, 1990.

Until 1984 the Bell operating companies and Western Electric were part of the same organization. This Western-operating company relationship was at the heart of the 1949 government antitrust suit and the subsequent 1956 consent decree. In the 1974 suit, less emphasis was given to the integration between equipment and telephone network services, but it was nevertheless an important issue. In its case, the government purported to show that AT&T had acted to restrict competition in equipment markets through its procurement practices that were managed by Western Electric.

That AT&T may indeed have shown a preference for Western Electric equipment may be discerned from the close relationship between AT&T's Long Lines' and operating companies' spending on plant and equipment and Western Electric sales. Between 1970 and 1982, Western accounted for between 63.3 percent and 69.7 percent of all AT&T plant and equipment expenditures (table 4-4). There was very little movement in this share despite the changes in the structure of the world telephone equipment industry and rather volatile exchange rates because Western Electric did not move production facilities abroad.

Immediately after divestiture, AT&T's share of the regional Bell operating companies' investment outlays began to fall. Between 1986 and 1988, AT&T Technologies' total sales—including computers and other products formerly not permitted under the 1956 consent decree—were only 58 percent of total outlays on plant and equipment by regional Bell operating companies and AT&T. In the ten years before divestiture, AT&T's equipment sales had averaged 72 percent of total AT&T capital expenditures. Had AT&T continued to enjoy the same share of the market for AT&T-regional Bell operating companies in 1986-88, its equipment sales would have averaged $3 billion more each year.

The reasons for this decline in AT&T Technologies' (Western Electric's) market share are many. First, Northern Telecom's success in pioneering large digital switches led AT&T operating companies to begin to order them even before divestiture. Second, the newly divested regional Bell operating companies undoubtedly sought to diversify procurement partly to reduce their risk of being dependent on AT&T, a competitor in some markets and a potential competitor in others. Third, the detariffing of CPE has created a competitive market for this equipment. The Bell operating companies can no longer be assured of a guaranteed market for AT&T-supplied

Table 4-4. Western Electric Sales versus AT&T Expenditures on Plant and Equipment
Millions of current dollars

Year	Western Electric: Total sales (1)	Western Electric: Sales to AT&T divisions (2)	AT&T plant and equipment outlays (3)	Western sales ÷ AT&T P&E outlays (1) ÷ (3)	Western's share of AT&T P&E outlays (2) ÷ (3)
1970	5,856	4,843	7,159	0.818	0.676
1971	6,045	4,987	7,564	0.799	0.659
1972	6,551	5,438	8,306	0.789	0.655
1973	7,037	6,085	9,322	0.755	0.653
1974	7,382	6,601	10,074	0.733	0.655
1975	6,590	6,016	9,329	0.706	0.645
1976	6,931	6,477	9,847	0.704	0.658
1977	8,135	7,708	11,566	0.703	0.666
1978	9,522	8,801	13,670	0.697	0.644
1979	10,964	10,031	15,837	0.692	0.633
1980	12,032	11,175	17,301	0.695	0.646
1981	13,008	12,142	18,098	0.719	0.671
1982	12,580	11,706	16,798	0.749	0.697
1983	11,155	...	14,127	0.790	...
1984	11,887[a]	...	16,859[b]	0.705	...
1985	12,180[a]	...	19,379[b]	0.629	...
1986	10,586[a]	...	19,930[b]	0.531	...
1987	11,655[a]	...	18,968[b]	0.614	...
1988	11,409[a]	...	19,393[b]	0.588	...

Sources: Annual reports.
a. Product sales by AT&T; 1986–88 data are based on slightly different method than earlier years' data.
b. AT&T and seven regional Bell holding companies.

terminal equipment and need not necessarily purchase their equipment for resale from AT&T.

Recent Trends in the Market for Central Office Equipment

In the U.S. market most investment in central office switching equipment in the past few years has been in the form of digital (time- division) equipment to replace analog systems and to expand new long- distance systems. Between 1977 and 1983 the nominal value of U.S. shipments of central office equipment increased at a rate of only 5 percent a year, but since 1983 shipments have increased by more than 17 percent a year in current dollars. Much

Table 4-5. U.S. Digital Switch Market, by Lines Placed in Service, 1982–88
Percent

	1982	*1984*	*1985*	*1986*	*1988*
AT&T	<1	22	39	49	46
Northern Telecom	66	42	44	35	32
GTE[a]	3	28	13	10	21
Others	31	8	4	6	

Sources: For 1982 and 1986, Peter W. Huber, *The Geodesic Network: 1987 Report on Competition in the Telephone Industry* (Department of Justice, 1987), table CO-4. For 1984–85, Jacquie McNish, "Digital Switch Business Sees More Competition," *Wall Street Journal*, January 22, 1986, p. 6. For 1988, author's calculations based on Terry Sweeney, "Fujitsu Gear Pegged for Bell Testing," *Communications Week*, May 15, 1989, p. 5.

a. GTE has recently formed a joint venture with AT&T for the production of central office switches (Carol Wilson, "AT&T, GTE Kick Life into Their Switching Venture," *Telephony*, January 9, 1989, pp. 10-11).

of this increase was undoubtedly because of the investments required to meet the equal- access provisions of the modified final judgment, which appear to have leveled off since 1986. The older electromechanical switches could not be easily modified to provide equal access. Moreover, the new electronic digital switches can provide numerous new services, such as voice storage, that the operating companies are now promoting.

For years Western Electric and GTE dominated the U.S. central office market, largely because of their captive sales to their operating companies and Western's sales to Long Lines. In 1979 Western Electric had 85.0 percent of the market.[5] By 1984 the overall central office equipment market was very different as AT&T's share slipped to 46.8 percent while Northern Telecom's share rose to 32.6 percent and GTE's to 16.3 percent of lines shipped.[6] This huge gain by Northern Telecom may be traced to its early lead in the digital switch market, which currently accounts for about 75 percent of the total central office market.

Northern Telecom installed its first digital time-division switch in 1977, and in 1980, as already discussed, its local office switch was approved by the Bell System. By 1984 Northern Telecom's share of the digital switch market was 42 percent, nearly double AT&T's 22 percent share (table 4-5). Even GTE exceeded AT&T's

5. *Telecommunications in Transition: The Status of Competition in the Telecommunications Industry*, Committee Print, Subcommittee on Telecommunications, Consumer Protection, and Finance of the House Committee on Energy and Commerce, 97 Cong. 1 sess. (Government Printing Office, 1981), p. 181.

6. International Trade Administration, *A Competitive Assessment of the U.S. Digital Central Office Switch Industry* (Department of Commerce, September 1986), p. 23.

1984 market share, shipping 28 percent of the digital lines that year. By 1986, however, AT&T had strongly rebounded to claim 49 percent of the central office digital switch market for central offices.[7] Since 1984 AT&T and Northern Telecom have accounted for more than 80 percent of the market as Siemens, Stromberg-Carlson, NEC, CIT-Alcatel, and other suppliers struggle to establish themselves as "third" suppliers to the large telcos. Thus, one must conclude that divestiture has not yet had a disastrous impact on AT&T's share of the U.S. market for central office equipment. AT&T's loss of market share in telephone equipment must be in other equipment.

The development costs of large digital switches and the difficulty in selling them to national postal, telegraph, and telephone authorities (PTTs) has limited the number of world competitors. Among U.S. companies, only ITT, Stromberg-Carlson, and Digital Switch Corporation have attempted to compete with Western Electric, GTE, and the other chief international competitors, but the U.S. companies have encountered serious difficulties. ITT has recently withdrawn from the U.S. market and merged its telephone equipment operations into a French company, CIT-Alcatel. Stromberg-Carlson has been merged into the British firm, Plessey; Digital Switch has halted production on its Class 5 digital switch; and GTE has formed a joint venture with AT&T in the U.S. market.

These recent changes leave only the large national companies that have traditionally enjoyed favored status with their PTTs: Siemens (Germany), NEC (Japan), Ericsson (Sweden), Alcatel (France), Plessey (U.K.), Northern Telecom (Canada), and AT&T (United States). Remarkably, there has been little entry from other electronics firms, and the competitors for modern, sophisticated, time-division digital switches with stored program control are mostly the same companies that dominated the earlier electromechanical switch market.

The divestiture of Western Electric from the Bell operating companies and the increase in competition in interexchange markets might be expected to create greater incentives for innovation in switching technology. The threat of bypass, the growth of PBXs and local area networks, and pressures to reduce state control of entry into local markets may have placed great pressures on telephone

7. This rebound was aided by the sharp rise in CO lines shipped between 1985 and 1986, undoubtedly because of the regional Bell operating companies' response to the equal-access provisions of the modified final judgment.

operating companies that were not present before 1982. In the past four years, the divested Bell operating companies have begun to experiment with local (Class 5) digital switches provided by NEC, Ericsson, and Siemens. This additional potential competition appears to be driving switch prices down rather rapidly.

It is difficult to construct a price index for any rapidly changing durable good. Telephone switches have undergone enormous technological changes in the past two decades. Analog crossbar switches have been replaced by sophisticated, electronic, time-division digital switches. As technology has improved, new generations of switches replace older switches that were able to perform fewer functions.

Bellcore has attempted to measure the real cost of switching over successive generations of AT&T switches. Its analysis of the installed cost of medium and large local office switches shows that real switch prices fell by about 8 percent a year from 1972 through 1981, but rose briefly in 1981–82, before resuming their decline after 1982 (figure 4-1). More recently, New York Telephone has reported to the New York Public Service Commission that its real switch prices per line fell by 10.5 percent a year from 1983 through 1988.[8] Kenneth Flamm's analysis of smaller switches bought by companies subsidized by the Rural Electric Administration shows a much more rapid rate of progress between 1982 and 1985.[9] Thus, it seems that switching costs have fallen more rapidly since 1983 than before divestiture.

The rate of progress in telephone switching has been far slower than the progress in computer processors or office equipment. Since 1972 the real price of computers has fallen by more than 20 percent a year.[10] Given that telephone switching systems are essentially large computers, one might have expected similar progress in

8. Eli M. Noam, "An Overview of Telecommunications in the United States of America," New York State Public Service Commission, October 25, 1988, fig. 12, deflated by the producer price index for finished goods.

9. See Kenneth Flamm, "Technological Advance and Costs: Computers versus Communications," in Robert W. Crandall and Kenneth Flamm eds., *Changing the Rules: Technological Change, International Competition, and Regulation in Communications* (Brookings, 1989), pp. 13–61. Producer price indexes for switches are available only since December 1985. Surprisingly, the PPI for electronic switches, deflated by the PPI for finished products, rose by 3.2 percent a year between December 1985 and June 1988.

10. Bureau of Economic Analysis, price index for computer processors, deflated by the PPI for finished goods.

Figure 4-1. Real Switching Costs, 1972–87

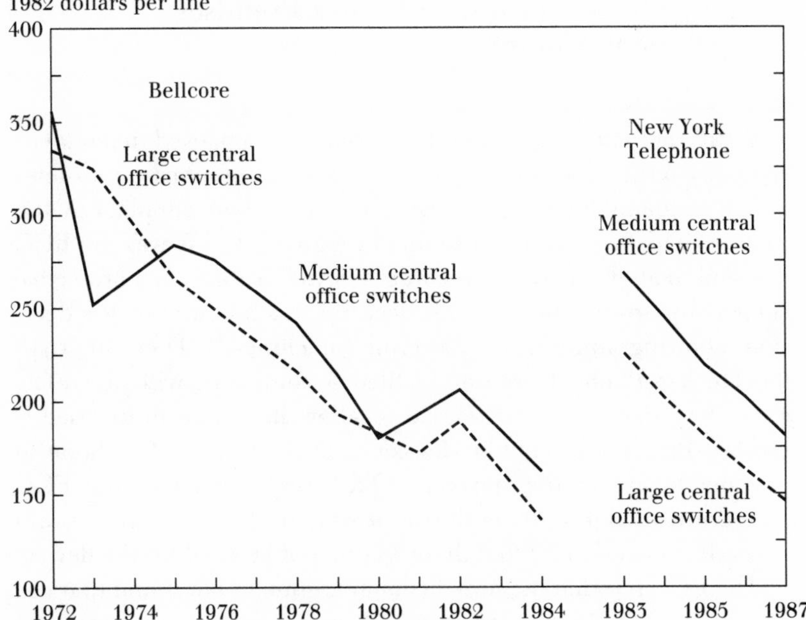

Sources: Kenneth Flamm, "Appendix B: Data," in Robert W. Crandall and Kenneth Flamm, eds., *Changing the Rules: Technological Change, International Competition, and Regulation in Communications* (Brookings, 1989), p. 404; and Eli W. Noam, "An Overview of Telecommunications in the United States of America," New York State Public Service Commission, October 25, 1988, fig. 12. Values have been deflated by the BLS producer price indexes for total finished goods.

telephone switching.[11] Unfortunately, the available indexes of telephone switching costs probably understate the rate of technical progress in switching because they do not capture the value of all of the service enhancements provided by the newer switches. However, it seems unlikely that even these omitted quality improvements could account for the threefold difference between the rate of progress in computers and the Bellcore estimates of progress in switching costs. Some part of the difference is undoubtedly because regulation has reduced the incentives for innovation.[12]

11. Bridger Mitchell reminds me that switches include line and trunk terminating equipment that have different progress curves.

12. Flamm, "Technological Advance." Switching costs fell sharply in the period after 1976 when competition accelerated in the PBX market. See table 4-6 and the discussion in the text.

Competition and the Customer Premises Equipment Market

Until the late 1940s the FCC had not opposed telephone-company tariffs that prohibited subscribers from attaching to telephone company lines any devices other than those supplied by the carriers. The first exception to these restrictive tariffs was the FCC decision to allow the use of recording devices subject to a protective connection arrangement.[13] This decision was followed by the decision allowing automatic answering machines.[14] Then in 1948 the Hush-a-Phone Corporation filed a complaint with the FCC requesting that tariffs be revised to allow its device to be used.[15] Hush-a-Phone sold cuplike devices that attached to the phone to provide privacy to the speaker. AT&T responded that the FCC lacked jurisdiction to issue the order sought, that the device would harm the network, and that there was no public need for the device.

The FCC rejected the jurisdictional argument but found that use of the Hush-a-Phone device could result in a deterioration of interstate service, and thus it would not be unreasonable to allow AT&T the right to prohibit its use.[16] The decision was appealed and the court found that the only parties affected were those using the device; therefore, the FCC could not exercise control over subscribers by allowing AT&T to restrict their right to converse in low tones.[17] The FCC then revised its decision to agree with the court's standard that a device could be used if it provided private benefit without public harm. AT&T refiled its tariffs; however, they were so narrowly defined that the only non-Western Electric equipment that could be used was the Hush-a-Phone device.

The next challenge to AT&T's restrictive tariffs came from the Carter Electric Company in the early 1960s. Carter made a device called a Carterfone, a cradle that would connect an ordinary

13. *Use of Recording Devices*, 11 FCC 1033 (1947).

14. *Jordaphone Corporation* v. *American Telephone and Telegraph Co.*, 18 FCC 644 (1954).

15. This controversy peaked in the late 1940s and 1950s, but it began in the 1920s.

16. *Hush-A-Phone Corporation*, 20 FCC 424 (1955).

17. *Hush-A-Phone Corporation* v. *United States and Federal Communications Commission*, 238 F. 2d 266 (1956).

telephone (acoustically) to a mobile radio transmitter. AT&T protested the use of the Carterfone, claiming that it violated its tariffs, leading Carter to sue AT&T for violating the antitrust laws. The district court in which the case was filed referred the matter to the FCC. In 1966 the commission initiated an investigation into the matter, and in 1968 issued a ruling that the AT&T tariffs excluding such devices were illegal.[18] AT&T had to refile its tariffs and in so doing specified that competitive devices could only be connected to the network through a telephone-company supplied device, called a connecting arrangement.[19]

These new tariffs were restrictive in a different way. In essence, a subscriber could attach a non-Western device, but would have to pay an extra price for the connecting device that was not required for Western Electric devices. AT&T was still restricting entry into the terminal equipment market.

Four years later, the FCC began an investigation into the need for connecting devices. After three years of proceedings, the requirement was repealed. In its place, a certification program was established for all equipment, including telco-supplied equipment.[20] As long as any equipment could be shown to be unharmful to the network and could be connected through standard plugs or jacks, it could now be used lawfully on the regulated network.

This certification program was challenged in the appellate courts, delaying its implementation until 1977. During this period, the North Carolina Utility Commission challenged the right of the FCC to force competition by preventing discrimination against subscribers who use non-AT&T terminal equipment. The court of appeals determined that the states' jurisdiction was limited to local services, facilities, and disputes that are separable from and do not seriously affect the conduct and development of interstate service.[21] With this decision, AT&T finally lost the ability to use state regulation to restrict the interconnection of non-Western Electric equipment.[22]

18. *Use of Carterfone Device*, 13 FCC 2d 420 (1968).

19. Gerald W. Brock, *The Telecommunications Industry: The Dynamics of Market Structure* (Harvard University Press, 1981).

20. *Proposals for New or Revised Classes*, 56 FCC 2d 593 (1975).

21. *North Carolina Utilities Commission* v. *FCC*, 552 F. 2d 1036 (1977), *cert. denied*, 434 U.S. 874 (1977).

22. For an extensive review of this history, see Fred W. Henck and Bernard Strassburg, *A Slippery Slope: The Long Road to the Breakup of AT&T* (Greenwood Press, 1988), chap. 11.

As the AT&T tariffs restricting interconnection were being challenged in the FCC in the 1940s and the 1950s, the Justice Department was pressing its 1949 antitrust suit against AT&T. The government charged Western with the unlawful control of virtually the entire market for telecommunications equipment and apparatus in the United States, elimination of competing manufacturers, and coordination of the operations of the Bell System so as to control equipment prices. The Justice Department initially sought relief that would break Western Electric into three companies, require Western to divest itself of its share of Bell Laboratories, and require Western to refrain from entering exclusive contracts with the Bell System for the provision of equipment. The complaint was resolved by the 1956 consent decree, which allowed AT&T to keep Western Electric but restricted AT&T to the manufacture of equipment and the provision of services related to regulated telecommunications common carriage.[23]

The distinction between communications and data processing or transformation began to blur in the 1960s and early 1970s. By 1965 the changing nature of telecommunications services prompted the FCC to begin what became known as the First Computer Inquiry. This inquiry was to look into the convergence of computer and communications technologies. In these proceedings, the FCC attempted to distinguish between regulated communications services and unregulated data processing. The commission concluded that a communications service was one in which data-processing capability was incidental to the message-switching function. A data-processing service, however, was defined as one in which computing rather than transmission was dominant. To prevent carriers from cross subsidizing between the two definitionally different services, the FCC ruled that communications carriers providing data-processing services needed to do so through a separate subsidiary.[24] This decision was appealed, but the court of appeals reaffirmed the separate subsidiary requirement. The court further interpreted the decision in conjunction with the 1956 consent decree as precluding the Bell System from offering data processing.[25]

23. Consent Decree, *United States* v. *Western Electric Co.*, 1956 Trade Cases (CCH) para. 68, 246 (D.N.J. January 24, 1956).

24. *Computer I*, 28 FCC 2d 267 (1971).

25. *GTE Services Corporation* v. *Federal Communications Commission*, 474 F. 2d 724 (1973).

When the smart terminal was introduced, difficulties arose with the *Computer I* decision. These terminals did not function as computers but performed some data conversions that allowed more efficient transmission. Thus, the definition of data processing became further blurred, and the FCC responded by reopening the Computer Inquiry.

The tentative decision in the Second Computer Inquiry was an attempt to define basic telecommunications equipment more narrowly. Basic equipment was defined as those devices that lacked intelligence. This classification of equipment came under extreme criticism and was dropped in the final decision. *Computer II* resolved this definitional problem by deregulating all equipment on customers' premises, whether intelligent or not.

Computer II also required that AT&T and GTE offer CPE through separate subsidiaries. In addition, the second inquiry suggested that the subsidiary of AT&T could offer enhanced equipment without violating the 1956 decree. *Computer II* was designed to limit common carrier regulation to basic transmission services. *Computer II* also took terminal equipment from carrier services and deregulated it. Finally, the FCC decided in *Computer II* to resolve the details of deregulating or "detariffing" CPE in separate proceedings.[26]

The issues of detariffing and the provision of customer premises equipment took a new turn with the announcement of the modified final judgment (MFJ) in 1982, settling the Justice Department's 1974 antitrust suit brought against AT&T. The MFJ contained several key provisions concerning CPE. First, at divestiture, all embedded CPE was to be transferred from the operating companies to AT&T. Furthermore, the divested Bell operating companies were restricted to marketing CPE; they were forbidden to manufacture it. The MFJ put pressure on the FCC to consummate its plan to transfer and detariff embedded CPE by January 1, 1984.

Computer II had previously established that existing or embedded CPE would be defined as any CPE owned by a carrier and subject to tariffs or to the separations process by January 1, 1983. New CPE, any CPE that went into use after that date, was immediately detariffed to allow for greater competition in the equipment market.

The transfer of CPE from the Bell operating companies to AT&T

26. *Second Computer Inquiry*, 77 FCC 2d 384 (1980).

Table 4-6. U.S. PBX Market Shares, by Lines Shipped, 1970–89
Percent

Company	*1970*	*1976*	*1978*	*1980*	*1982*	*1984*	*1986*	*1988*	*1989*
AT&T	80	50	48	46	23	20	22	22	28
NorthernTel	...	13	9	8	13	20	23	19	23
Rolm	...	4	7	10	13	14	16	16	12
Mitel	...	0	0	7	12	10	10	8	9
NEC	...	5	3	3	5	8	8	8	7
GTE	6	1	7	6	5	5	4	...	...
Siemens	...	0	1	4	5	4	4	5	5
Others	14[a]	27	25	16	24	19	13	22	16

Sources: For 1970, *Telecommunications in Transition: The Status of Competition in the Telecommunications Industry*, Committee Print, Subcommittee on Telecommunications Consumer Protection and Finance of the House Committee on Energy and Commerce, 97 Cong. 1 sess. (Government Printing Office, 1981), p. 184; for 1976–82, Richard H. K. Vietor and Davis Dyer, eds. *Telecommunications in Transition* (Harvard University School of Business, 1986), p. 62; for 1984, Francis McInerney, "PBX Market Analyses," *Teleconnect*, September 1, 1985, p. 90; for 1986, P. Huber, *Geodesic Network table PX-4; for 1988, North American Telecommunications Association, Telecommunications Market Review and Forecast: Annual Report of the Telecommunications Industry, 1990 edition* (Washington, 1989), p. 120; and for 1989, *Communications Week*, January 8, 1990, p. 28.

a. Includes all U.S. and foreign suppliers except GTE and AT&T.

required the establishment of a fair value to transfer the equipment and the development of a sale and lease program that would best serve the subscribers. The embedded base was to be transferred at its net book value. Sale and lease programs for single and multiline equipment were carefully prescribed to protect the customers. The independent telephone companies were addressed separately. They were required to detariff their CPE by December 31, 1987.

Private Branch Exchanges

The private branch exchange (PBX) is a small switching system, similar to a central office switch at a telephone company, but located on the customer's premises. Since the deregulation of PBX systems in the late 1970s, foreign suppliers have entered the U.S. market in force. Total shipments of PBXs increased dramatically through 1982 but have flattened out considerably since then (table 4-3), probably because of the resurgence of Centrex service and the failure of PBXs to penetrate the market for data communications.

U.S. manufacturers began losing market share to companies such as Northern Telecom, Siemens, and NEC in the 1970s (table 4-6), but recent data suggest that the erosion in U.S. producers' share has been arrested by the 1985–87 decline in the value of the dollar.

AT&T's share of the PBX market began falling in the early

1970s. After the resolution of the issue concerning network protective devices in 1978, this decline accelerated. By 1984 AT&T had barely one-fifth of the market. Thus, even before divestiture, the PBX market had become quite competitive, and AT&T enjoyed little of the market power that it had only a decade earlier. It was not divestiture, but earlier decisions by the FCC to allow interconnection of competitive equipment that eroded AT&T's market power.

Despite the volatile nature of individual producers' market shares, the PBX market remains modestly concentrated. In 1984, for example, the top five producers accounted for 71 percent of the market;[27] by 1989 they had increased their share to 79 percent. But the many large international firms now competing in this market limit the ability of U.S. firms to restrict output and raise prices in the domestic market.

Key and Hybrid Telephone Systems

A large increase in shipments of key telephone systems took place in 1984 as detariffing of CPE and divestiture occurred. Since then, however, competition from small PBXs, Centrex, and imported systems has placed strong downward pressure on U.S. shipments.[28]

At the beginning of 1982 AT&T held almost a 70 percent share of the market for key telephone systems. By 1988 AT&T's share had fallen to 25 percent, and its top competitor, TIE, had about 17 percent of the market.[29] TIE has been particularly successful at getting contracts with the divested Bell operating companies, who now account for nearly 20 percent of distribution.[30]

Personal Phones

In the early 1980s, the personal telephone market expanded at a very rapid rate as consumers replaced their telco-owned instruments with their own telephones. In the late 1980s, however, this growth

27. "New Waves in PBX Market," *Telephone Engineer and Management*, November 1, 1986, pp. 58–59.

28. North American Telecommunications Association (NATA), *Telecommunications Market Review and Forecast: Annual Report of the Telecommunications Industry* (Washington, 1989).

29. NATA, *Telecommunications Market Review*.

30. Peter W. Huber, *The Geodesic Network: 1987 Report on Competition in the Telephone Industry* (Department of Justice, 1987), table T-7.

began to moderate.[31] Possibly, the demand for personal telephones will begin to rebound as consumers replace some of the low-quality instruments sold in the early 1980s.[32]

AT&T remains the chief supplier of telephone handsets in the United States, accounting for 36 percent of shipments in 1986, but its share has fallen by half since the FCC implemented its competitive equipment certification program in the late 1970s.[33] No other manufacturer has as much as 10 percent of the market. Moreover, most handsets are now manufactured abroad (table 4-2); even AT&T has shifted much of its production to lower-wage countries in Asia.

Call/Voice Handling Equipment

Perhaps the fastest growing terminal equipment market is that for call/voice handling equipment. The proliferation of PBXs in the 1970s and 1980s allowed users to add automatic call distributors to route calls to the appropriate operator or office. Voice messaging equipment has also grown rapidly as the demand for automated voice-mail services has increased. As table 4-7 shows, sales of these various types of equipment have been growing more rapidly than any other category of terminal equipment.

Data-Communications Equipment

The slow growth in the PBX and key systems markets reflects the limited potential for growth of traditional voice traffic. Data traffic between computers is growing much more rapidly, and much of this traffic is being routed through private local-area networks. As a result, sales of devices such as network controllers are growing rapidly. The sales of all principal components used in data-communications systems have now surpassed the revenues derived from traditional key telephone systems and PBXs (table 4-7) and are growing far more rapidly.

31. NATA, *Telecommunications Market Review*, p. 154.

32. Andrew Pollock, "The Phone Supplier Shakeout: Gain in Sales Not Enough," *New York Times*, March 27, 1984, p. D1.

33. Huber, *Geodesic Network*, table T-5; and *Telecommunications in Transition*, Committee Print, p. 181.

Table 4-7. Growth of U.S. Terminal Equipment Markets, 1984–88
Millions of current dollars

Type of market	*1984*	*1985*	*1986*	*1987*	*1988*	*Average annual growth rate (percent)*
Consumer equipment						
Telephone sets-corded	1,200	1,585	1,685	1,750	1,825	10.5
Telephone sets-cordless	410	305	410	438	474	3.6
Answering equipment	298	371	535	557	634	18.9
Business equipment						
PBX	2,843	3,086	3,105	3,050	2,928	0.7
Key systems	1,039	955	957	960	965	−1.8
Data communications						
Modems	1,100	1,266	1,500	1,650	1,840	12.9
Multiplexers	550	613	700	840	910	12.6
Communications processors	751	790	815	845	896	4.4
Concentrators	165	180	197	209	219	7.1
Local area networks	300	450	700	840	1,200	34.7
Other (including packet switching)	325	455	620	742	961	27.1
Facsimile equipment	175	164	260	821	1,825	58.6
Cellular equipment	75	280	393	475	675	54.9
Call/voice handling equipment						
Voice messaging equipment	5	75	150	315	500	115.1
Automatic call distributors	95	183	228	285	350	32.6
Voice-processing equipment	10	35	75	160	235	78.9
Other equipment	2	15	30	70	100	97.8

Source: North American Telecommunications Association, *Telecommunications Market Review and Forecast*, 1990 edition, pp. 12, 144, 149, 154, 162, 178.

Facsimile

At the time of divestiture, the facsimile market was relatively stagnant with annual sales of only about $175 million. Between 1986 and 1987, however, the market began to expand rapidly as average prices for terminals fell from the $4,000 range to slightly over $2,000. By 1988 sales for facsimile terminals were $1.8 billion, and they were expected to double by 1990.[34]

Cellular Telephones

Total cellular equipment sales have increased steadily since 1984, but this technology is still in its infancy. By 1988 there were nearly

34. NATA, *Telecommunications Market Review*, table 43.

2 million subscribers, and new sales of cellular telephones grew to $675 million. Unit sales are expected to exceed 1.75 million in 1990 as the number of subscribers increases to 4 million.[35] As with facsimile machines, much of this growth has been stimulated by big reductions in price from $2,000 to $4,000 in 1985 to as little as $300 to $500 in 1990.

Terminal Equipment Prices

The only consistent source of data on terminal equipment prices before 1985 is the AT&T telephone plant index (TPI), an internal index produced for the Bell System for planning and regulatory purposes. Unfortunately, as Flamm has argued, this index suffers from its inability to reflect the improvements in quality from one generation of equipment to another.[36] Thus, the TPI probably overstates the rate of price increase for most telephone equipment.

The TPIs for large PBXs, key sets, and telephone sets for 1972 and 1985 are shown in figure 4-2. Throughout the early 1970s, real prices fell rapidly for key sets and ordinary handsets, but the real price reductions for PBXs were more modest. This decline slowed dramatically between 1975 and 1977 but then accelerated once more in 1978. During this period the FCC was fighting off legal attempts to block its competitive registration program. When all legal challenges were finally turned aside in 1978 and the FCC adopted its final rule, the prices of terminal equipment began to fall once more. Between 1972 and 1981, the average rate of real price decline for these three types of equipment was about 6 percent a year.

Between 1981 and 1983, the real TPIs for these items reversed once more, especially for key sets. During those years AT&T was preparing for divestiture. In 1982 and 1983 the TPIs reflected transfer prices within the Bell System. After 1983 the TPIs reflect the prices paid by New York Telephone for equipment purchased at arm's length.

Between 1983 and 1987, the real prices of PBXs and key/hybrid telephone sets fell once again by about 6 percent to 7 percent a year, or at about the same rate as in 1972–81 (figure 4-3), according

35. NATA, *Telecommunications Market Review*, p. 178.
36. Flamm, "Technological Advance."

Figure 4-2. Real Price Trends for PBXs, Telephone Sets, and Key Sets, 1972–85
Price per line in constant dollars, 1977 = 100

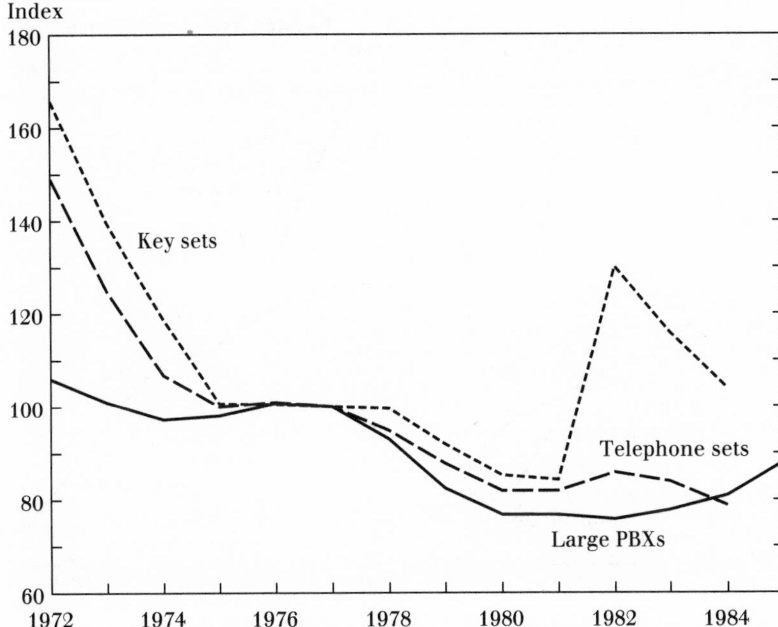

Sources: For 1972–82, AT&T, "Bell System Telephone Plant Indexes," SL83-05-011, memorandum, New York, May 31, 1983; for 1983–85, Bell telephone plant indexes and Nynex, "Telephone Plant Index," memorandum, White Plains, N.Y., January 1987.

to industry data. Of course, the rate of real price improvement is understated by all of these data, given the improvements in technology that are occurring in the postdivestiture world.

Employment, Productivity, and Wages

The new entry into the U.S. telephone equipment industry and the shift of demand from traditional telephone technology to the newer fiber optics, satellite, and radio-based technologies could be expected to create great turmoil in the industry. Moreover, one would expect that competitive entry and divestiture would have created pressures for improvements in productive efficiency. The trends in employment in the 1980s confirms this prediction. Employment in the telephone and telegraph apparatus industry reached its

Figure 4-3. Real Price Trends for PBXs and Key and Hybrid Telephones, 1982–89
Price per line in constant dollars, 1985 = 100

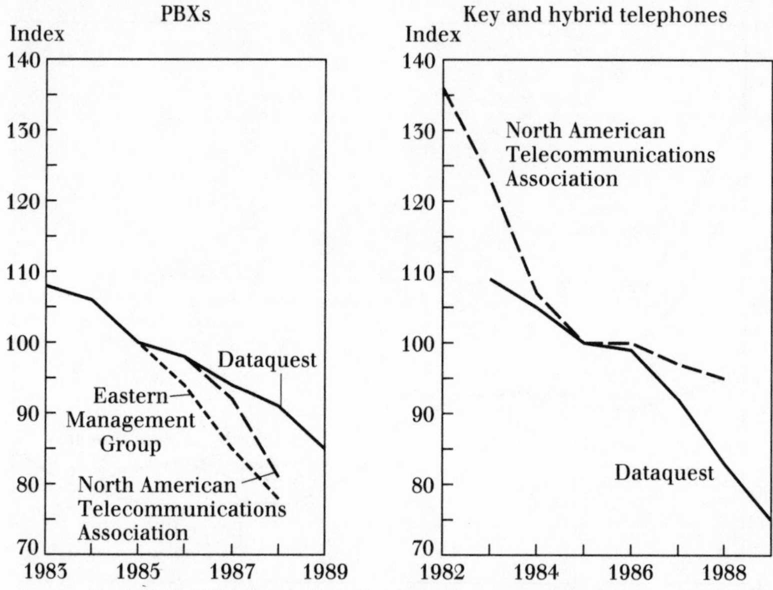

Sources: For 1982–83 NATA figures, North American Telecommunications Association (NATA), *1987 Statistical Review: Annual Market Study of the Telecommunications Equipment Industry* (Washington, 1987), p. 16; for 1984–88, NATA, *Telecommunications Market Review and Forecast: Annual Report of the Telecommunications Industry* (Washington, 1989), pp. 118, 133. For 1983–87 Dataquest figures, Dataquest Inc., "Datafacts," San Jose, Calif., May 1988, November 1988, app. A and table A-1, respectively. For 1988–89, David Falgetter, Dataquest Inc. For Eastern Management Group figures, Mark Ricca, Eastern Management Group, Parsippany, N. J., February 1989. Indexes have been deflated by BLS producer price index for total finished goods.

peak in 1980. In 1981 employment began to decline, and this decline continued through 1989 (table 4-8), exacerbated by declining demand for traditional telephone equipment.

In the radio communications equipment industry (3662), employment continued to rise through 1986, but has fallen since then (table 4-9). Much of the increase in employment in SIC 3662, however, occurred because of the sharp rise in defense spending from 1981 to 1985, and the slowdown since 1985 is also because of a decline in defense-related output.

Since 1977 the average annual rate of growth in output per employee hour in the telephone and telegraph equipment industry has remained about 5 percent (table 4-7), but this result is heavily dependent on estimates of industry revenues and a deflator that may be biased upward over time. One simply cannot tell from current data if divestiture has seriously affected productivity growth.

Table 4-8. Employment Output per Employee Hour and Value of Shipments for the Telephone and Telegraph Apparatus Industry (SIC 3661), 1975–89

Year	*All employees (thousands)* *(1)*	*Production workers (thousands)* *(2)*	*Output per employee hour (1977 = 100)* *(3)*	*Real value of shipments (1977 = 100)* *(4)*
1975	147.5	94.4	81.0	79.6
1976	137.2	89.6	89.3	74.7
1977	147.0	99.4	100.0	100.0
1978	155.0	106.5	105.4	113.3
1979	165.4	115.8	115.1	133.9
1980	163.7	112.1	116.6	143.9
1981	157.5	104.8	124.7	147.1
1982	149.3	96.4	129.8	136.7
1983	140.6	88.8	138.1	133.2
1984	151.1	97.5	153.5	163.8
1985	139.9	88.5	161.1	179.1
1986	121.2	70.0	162.5	165.7[a]
1987	116.1	67.0	n.a.	166.0[a]
1988	117.1	69.3	n.a.	167.9[a]
1989	102.8	61.1	n.a.	n.a.

Sources: For columns 1 and 2, 1975–1976, Bureau of Labor Statistics (BLS), *Employment and Earnings, United States, 1909–78* (Department of Labor, 1979); for 1977–86, BLS, *Supplement to Employment and Earnings Revised Establishment Data* (Department of Labor, annual issues); and for 1987–89, BLS, *Employment and Earnings* (Department of Labor, annual March issues), table B-2. For column 3, BLS, SIC 3661 Telephone and Telegraph Apparatus Output per Employee Hour, data base (February 1989). (Hereafter BLS Radio and TV data base.) For column 4, value of shipments, Bureau of the Census, "Selected Electronic and Associated Products, 1981," series MA36N, and "Communications Equipment, 1986, 1987, 1988," series MA36P. Deflated by BEA output index for SIC 3661 obtained from John Musgrave, BEA. All data are for the pre-1987 definition of SIC 3661.

n.a. Not available.

a. The 1986–88 data are not strictly comparable with earlier data owing to revisions in SIC industry classifications.

Productivity growth found in SIC 3662, radio communications equipment, has been slower, but the large share of defense spending in this industry makes it more difficult to compare its performance with that of the telephone equipment industry.

Average hourly earnings in the telephone and telegraph equipment industry rose steadily between the mid-1970s and the early 1980s. However, the rate of growth began to decline in 1982, and by 1985 average real wages were no longer rising (table 4-10). Apparently, increased competition in the product market and a shift of demand away from AT&T have led to a downward pressure on real wages since 1982, a result that has become standard in deregulated industries.

Table 4-9. Employment and Output per Employee in Radio Communications Equipment (SIC 3662), 1975–89

Year	*All employees (thousands)* (1)	*Production workers (thousands)* (2)	*Output per employee hour* (3)
1975	309.8	126.2	90.4
1976	308.3	124.1	94.2
1977	314.6	125.5	100.0
1978	337.1	134.3	102.6
1979	357.3	146.4	113.7
1980	377.7	154.6	121.5
1981	399.3	165.3	121.0
1982	420.1	169.3	122.4
1983	431.9	171.8	124.2
1984	463.7	185.2	130.5
1985	519.4	204.1	139.7
1986	537.1	202.8	144.1
1987	498.5	189.8	n.a.
1988	480.0	180.6	n.a.
1989	435.7	156.4	n.a.

Sources: For column 1 and 2, see table 4-8. For column 3, BLS Radio and TV data base. All data are for the pre-1987 definition of SIC 3662.
n.a. Not available.

Imports, Exports, and International Competition

The combination of divestiture and the high value of the dollar proved disastrous for the U.S. trade balance in telecommunications equipment. During the 1970s, the United States typically realized a small surplus in telephone equipment trade. This surplus continued through 1982, the year of the announcement of the modified final judgment (the 1982 decree) that divested Western Electric from the Bell operating companies. From 1982 through 1987, the U.S. trade balance in telephone equipment shifted from a surplus to a deficit of $2.7 billion, undoubtedly spurred by the strong dollar.[37] But the timing of the dramatic shift (1983) suggests that divestiture must have played some role. Had the Bell operating companies still been tied to Western Electric, they would probably not have moved

37 Bureau of the Cenus, "Communications Equipment including Telephone, Telegraph, and Other Electronic Systems and Equipment, 1987," *Current Industrial Reports*, series MA36P (Department of Commerce, 1988), table 4.

Table 4-10. Average Hourly Earnings for Production Workers in Telephone and Telegraph Equipment, Radio Communications Equipment, and All Manufacturing, 1972–89

	Telephone equipment (SIC 3661)		*Radio communications equipment (SIC 3662)*		*All manufacturing*	
Year	*Current dollars*	*1972 dollars*	*Current dollars*	*1972 dollars*	*Current dollars*	*1972 dollars*
1972	4.18	4.18	4.20	4.20	3.82	3.82
1973	4.45	4.19	4.41	4.15	4.09	3.85
1974	4.73	4.01	4.82	4.09	4.42	3.75
1975	5.23	4.06	5.26	4.09	4.83	3.75
1976	5.78	4.25	5.69	4.18	5.22	3.84
1977	6.28	4.34	6.14	4.24	5.68	3.92
1978	6.73	4.32	6.62	4.24	6.17	3.96
1979	7.50	4.32	7.13	4.11	6.70	3.86
1980	8.24	4.18	7.70	3.91	7.27	3.69
1981	9.15	4.21	8.52	3.92	7.99	3.68
1982	9.86	4.27	9.48	4.11	8.49	3.68
1983	10.27	4.31	10.28	4.32	8.83	3.71
1984	10.69	4.30	10.84	4.36	9.19	3.70
1985	10.82	4.21	11.44	4.45	9.54	3.71
1986	10.94	4.17	11.42	4.36	9.73	3.71
1987	11.17	4.11	11.96	4.40	9.91	3.65
1988	11.62	4.11	12.26	4.33	10.17	3.59
1989	11.68	3.94	12.73	4.29	10.47	3.53

Sources: For 1972–78, BLS, *Employment and Earnings, United States, 1909–78* (Department of Labor July 1979). For 1979–86, *Supplement to Employment and Earnings Revised Establishment Data* for 1987–89, *Employment and Earnings*, annual March issues, table C-2. The 1972 dollar values are current year values deflated by the consumer price index, all urban consumers for all items.

aggressively to purchase foreign equipment regardless of the level of exchange rates.

Much of the shift in the telephone equipment trade balance resulted from the deregulation and detariffing of CPE. A $1.3 billion deficit occurred in telephone sets alone in 1988 (see table 4-2). Thus, telephone sets accounted for half of the 1988 deficit even though such equipment accounts for only one-eighth of traditional telephone equipment consumption.

Often neglected in the discussion of telecommunications equipment trade is the generally favorable U.S. balance in nonbroadcast communications systems.[38] Throughout the 1980s, the U.S. trade balance has been positive but declining in these high-technology

38. Formerly SIC 36621, now SIC 36631.

equipment markets. Exports of this equipment have been much greater than exports of the traditional telephone and telegraph products. Although imports have risen since divestiture, U.S. exports of nonbroadcast communications systems have also risen greatly. Thus, divestiture's principal impact on the U.S. trade balance may have been simply associated with the detariffing of CPE as the Bell operating companies were divorced from AT&T.

The U.S. producers retain a very large share of the market for switching and transmission equipment despite divestiture. As the dollar recedes and the initial surge in consumer purchases of CPE ebbs, the trade deficit in telephone equipment will probably begin to decline. But even if it does not, the increase in market competition occasioned by the detariffing of CPE and the AT&T divestiture seems to have significantly improved U.S. economic welfare.

Real Price Changes in Telecommunications Equipment and Similar Products, 1960–86

The only available price series for telephone equipment have historically been the Bell System TPIs. These indexes do not capture the full effects of technical change, thus presenting a serious problem in analyzing the impacts of liberalization and divestiture on the telephone equipment industry. However, the available information on prices for telecommunications equipment can be compared with price data for similar products in other sectors of the economy.

Throughout the 1970s, the real value of the Bureau of Economic Analysis deflator for telecommunications equipment declined at a 3.5 percent annual rate (table 4-11). Similarly, the AT&T index for inside plant—the cost of installing central office equipment, consumer station connections, and customer premises equipment—fell at a 2.1 percent real rate. The Bellcore estimate of real switch prices for large offices fell at a 5.7 percent rate in the 1970s. The AT&T index for outside plant, which includes more construction labor than the inside plant index, rose slightly in real terms in the 1970s.

In the 1980s, these telecommunications price indexes began to rise in real terms. Figure 4-1, however, shows that prices for central office equipment continued to decline in the 1980s, and figure 4-2 shows a sizable decline in the real prices of PBXs and key systems.

Table 4-11. Annual Percent Changes in Real Prices of Telecommunications Equipment and Similar Products, Selected Periods, 1960–86

Item	*1960–69*	*1970–79*	*1980–86*
Telecommunications equipment			
BEA deflator for telecommunications equipment	−0.04	−3.5	2.8
AT&T telephone plant index			
Inside plant	1.9	−2.1	2.0[a]
Outside plant	3.0	0.5	0.4[a]
Bellcore central office switching (large offices)	n.a.	−7.4[b]	n.a.
Similar products			
BEA computer systems	n.a.	−23.5	−19.4
PPI electronic components	n.a.	−3.7	1.6
PPI nonferrous wire cable	2.9	−2.7	−3.6
PPI home electronics equipment	−4.0	−8.3	−5.1

Sources: BLS, Labstat Series Report: Producer Price Indexes, data base (January 1988); Kenneth Flamm, "Appendix B: Data," in Robert W. Crandall and Kenneth Flamm, eds., *Changing the Rules: Technological Change, International Competition, and Regulation in Communications* (Brookings, 1989), p. 404; and New York Telephone, "Telephone Plant Indexes," White Plains, N.Y., January 1987. BEA telecommunications equipment deflator obtained from John Musgrave, BEA; and BEA computer systems obtained from David Cartwright, BEA. All series deflated by the producer price index for finished goods.

n.a. Not available.

a. 1980–85.

b. 1972–81.

Why the AT&T and BEA indices should show a rise in the 1980s, therefore, is a puzzle.

The rate of decline in the real prices of computers and other office machines has been far greater than the real decline in prices of telephone central office equipment since 1970. In the 1980s, the rate of price decline in computers continued to be near 20 percent a year. Throughout the 1970s and 1980s, the price performance of home electronic equipment, nonferrous wire, and electronic components was far superior to the performance in telecommunications equipment, although the rate of price improvement in most terminal equipment was about the same as in the producer price index for home electronics equipment.

Given the intensity of the competition in CPE and the obvious continuing technical progress in switching equipment, the Bureau of Economic Analysis and AT&T indexes of telephone equipment cost have undoubtedly failed to capture all of the real price declines inherent in new generations of switching systems, transmission systems, and terminal equipment. If this is true, it is likely that I have underestimated telephone capital formation as reported in table 3-4 and overestimated productivity growth in telephone services, but underestimated labor productivity growth in the telephone equipment industry.

A Concluding Assessment

Clearly, CPE deregulation, the AT&T divestiture, and the strong dollar between 1980 and 1985 have combined to make the U.S. telecommunications equipment market much more competitive than it was in the 1960s and 1970s. The terminal equipment market, in particular, is much less concentrated today, and market shares of the principal terminal equipment producers are extremely volatile. AT&T's share of the PBX market fell dramatically before divestiture but has risen since then despite the effects of the strong dollar.

In the market for central office equipment, market shares change much more slowly. The choice of large switching equipment requires substantial lead times because of the complexities of network design. In the late 1970s Western Electric began to fall behind Northern Telecom in designing and producing digital switches for local exchange networks. Since the late 1970s, AT&T has lost a large share of the market for central office equipment, but it appears that it has recovered some of this share in the important digital switch business. Nevertheless divestiture has undoubtedly improved the access of other manufacturers to the Bell operating companies and has reduced AT&T's share of the total market. Divestiture also seems to have placed substantial downward pressure on the prices of switches and PBXs.

The most stunning effect of competition and divestiture, however, has been on employment. Competition from imports and from new firms in the United States has jolted the telephone equipment industry (SIC 3661). Firms from the radio communications equipment and computer industry are increasingly competing in the telecommunications equipment market and forcing the traditional suppliers of telephone equipment to improve productivity. Real output of U.S.–produced nonbroadcast communications equipment has risen greatly since 1982, but employment in this industry has fallen.

Unfortunately, the only data on predivestiture market performance are derived from internal AT&T transfer prices; hence, it is difficult to conclude that the postdivestiture market performance is superior to that of the pre-1984 years. Moreover, changes in the market for central office equipment may require a decade or more to analyze properly. The sharp rise in new types of terminal equipment—for facsimile, call/voice handling, and data communications—in recent years may not have occurred as rapidly in the predivestiture, preliberalization era; however, it is impossible to separate the impact of changes in market structure and competition policy from those of technology in the 1980s.

Chapter Five

Income Distribution and Economic Welfare

THE LIBERALIZATION of telephone services and equipment markets and the AT&T divestiture have placed enormous pressures on regulators, telephone companies, and labor. The entire telephone rate structure, a reflection of generations of political decisionmaking, is being radically revamped by regulators who are now much more attuned to market forces. Carriers must respond to the changes in regulation and to the rapid growth in competition. These responses, in turn, affect the efficiency of the entire telephone network.

In this chapter I analyze the distributional and efficiency consequences of these changes. In particular, I am concerned with the direct and indirect effects of repricing on the cost of telephone service to households by income class. I also estimate the overall welfare effects of repricing, liberalization, and divestiture.

The Impact of Competitive Entry and Divestiture on Telephone Rates, Quality of Service, and Income Distribution

Just how much have competition, divestiture, and regulation affected telephone rates, universal service, telephone subscription among lower-income households and rural residents, and the quality of service over the past twenty-five years? To begin with, many people fear that deregulation, competition, and divestiture have raised telephone rates, thereby reducing the ability of poorer households to afford basic service. As some of these households drop off the telephone network, imperiling universal service, the value of telephone service to others is reduced. If this scenario is accurate, competition and divestiture may have given the economy improved efficiency in the delivery and pricing of some services at the cost of reducing welfare through the loss of consumption externalities and adversely redistributing income.

The regulatory decisions to reprice telephone service may also have caused rural rates to rise. The reduction of subsidies flowing from long distance and from urban consumers to rural consumers may improve economic welfare generally but may also create political opposition to the liberalization of the telephone sector.

Telephone rates began to rise in real terms in the early 1980s before the AT&T divestiture. Since 1986 real rates have receded once again. In this section, I explore the reasons for this pattern of rate movements, looking especially for evidence that rate rebalancing or the AT&T divestiture may be to blame. I also review the recent data on quality of telephone services, looking for evidence that divestiture has been associated with a decline in the quality of telephone service available to subscribers.

The Effect of Higher Local Rates on Low-Income Subscribers

Undoubtedly, the regulatory pressures of the 1980s, which resulted from the FCC decisions of the 1970s, have led to a relative increase in local access rates and a relative decline in interexchange rates. This repricing of service has led to great concern that low-income households may be forced off the telephone network because of higher basic subscriber rates. In many jurisdictions, low-cost Lifeline rates have been developed to counter this possibility, subsidized in part by FCC programs that draw on the interstate carriers' access charges.

Because lower-income households are likely to make fewer toll calls, it is argued, these households bear the brunt of repricing of service. Even if these lower-income households continue to subscribe, the shift in the burden of non-traffic-sensitive costs to the individual subscriber may adversely affect the distribution of income.

Telephone access is clearly a necessity in that the demand for service is very price inelastic and only modestly income elastic. Estimates of the price elasticity of demand for access to telephone service usually range between −0.05 and −0.15.[1] Data from the Bureau of Labor Statistics (BLS) 1988 Consumer Expenditure Survey demonstrate that telephone expenditures vary inversely with income. Households in the lowest-income classes spent an average of between $352 and $368 a year, or about 3.2 percent of their total annual expenditures, on telephone service. Households with incomes of more than $40,000 spent less than twice as much each year on average as those with incomes of less than $5,000 a year, or about 1.2 percent of their total expenditures.

Critical to the debate on repricing, however, is the division of expenditures among local access, local usage, toll service, and various enhanced features. Studies of these divisions for lower-income subscribers show that lower-income subscribers spent about $7.00 to $25.00 a month in the mid-1980s for local service and between $17.00 and $21.00 a month on toll service. A 1986 study of low-income Michigan consumers found that telephone subscribers with an average income of $8,746 a year spent a monthly average of $25 on local service and $21 on long-distance calls.[2] A 1985 study of Pacific Bell subscribers found that those with annual incomes of less than $11,000 spent $15.00 a month on local service, but those with Lifeline services averaged only $7.00 a month on local services. A study by Southwestern Bell of low-income census tracts found that the average subscriber in these areas spent $15.63 a month on local services in the first half of 1988.[3]

In each of these studies, low-income subscribers were found to spend far more than the lowest monthly flat-rate charge available for access and local service. Even the lowest-income subscribers utilize certain custom features such as touch-tone service or call-

1. Lester D. Taylor, *Telecommunications Demand: A Survey and Critique* (Ballinger, 1980), chap. 3; and Lewis J. Perl, "Residential Demand for Telephone Service, 1983," National Economic Research Associates, White Plains, New York, December 1983.

2. "Low Income Households in the Post-Divestiture Era: A Study of Telephone Subscribership and Use in Michigan," Michigan Citizens Lobby, October 1986, p. 114.

3. Field Research Corporation, "Residence Customer Usage and Demographic Characteristics Study," conducted for Pacific Bell, June 1985, pp. 2, 6, 20, 24. Local service charges are total Pacific Bell charges minus intra-LATA charges. See Alexander C. Larson and others, "The Effect of Subscriber Line Charges on Residential Telephone Bills," *Telecommunications Policy*, vol. 13 (December 1989), pp. 338–53.

waiting, and obviously, low-income households amass local usage charges. In the Michigan and Southwestern Bell studies, these additional charges were generally in the range of $2 to $3 a month for low-income households. The lowest flat rates available from Bell operating companies in each state at the end of 1987 varied greatly, but the median of the midpoint of each state's flat-rate charges was $12.74, or $152.88 a year.[4] This figure is equal to 46 percent of the BLS estimate of annual expenditures for telephone service for the lowest-income class in 1987.[5] Thus, it is fair to say that in 1987, only about one-half of the average low-income household's expenditures on telephone service were just to gain access to the network. The remaining charges were for enhanced features, local usage, and toll services.

Even among low-income subscribers, however, the distribution of long-distance charges is highly skewed. The Michigan study found that the average long-distance bill reported by low-income subscribers in its sample was $21.10 a month, but the median long-distance bill was only $10.00 a month. Thus, while the increase in local charges owing to the subscriber-line charges may be greatly offset by lower inter- LATA rates for some low-income consumers, this offset is highly skewed even among poorer telephone subscribers.[6]

The most consistent recent information on telephone subscription is obtained from the Current Population Survey (CPS). The CPS has surveyed households for telephone availability since 1983, finding that the share of U.S. households with telephone service has expanded every year to more than 93 percent. The FCC reports detailed breakdowns of the CPS data that seem to show little decline in subscription levels for the lower-income classes, but these data may be misleading because a variety of different categories of households may be combined in the lower-income categories, including families with several children, single young adults, and even college students from reasonably affluent families.

4. National Association of Regulatory Utility Commissioners (NARUC), *Bell Operating Companies' Exchange Service Telephone Rates* (Washington, 1987). These data exclude Lifeline service.

5. Bureau of Labor Statistics, Consumer Expenditure Survey, data base (1987), table 2.

6. Leland Johnson, the Rand Corporation, has suggested to me that lower toll rates also increase the frequency of incoming calls, increasing economic welfare for their recipients, poor and rich alike. (Personal correspondence, May 19, 1989.)

To get a better picture of the recent changes in telephone subscription by demographic groups, I begin by identifying a few relatively homogeneous household categories, both from urban areas—two persons with a white head of household; two persons with a black head of household; three persons with a white head; and three persons with a black head. The CPS sample is then broken down by income category within these groups (table 5-1).

It is clear that in March 1987, the share of each type of household with telephone service rose with income at a declining rate. Black households with the lowest incomes clearly have lower levels of telephone penetration than white households, but the differences narrow substantially at incomes of more than $15,000 a year.

In recent work, Lewis J. Perl has estimated an extremely detailed telephone access demand equation from 1980 census data (not the CPS).[7] I use his estimated equation to predict telephone subscription levels for each demographic category in table 5-1 at rates prevailing at the end of 1986. It is obvious that the Perl equation tends to overpredict telephone penetration as reported to the CPS for the lowest-income classes, perhaps because his equation is estimated for census data while table 5-1 reflects a smaller sample that is disproportionately urban in character. However, the Perl equation predicts well for income classes above $15,000.

To measure the effect of local rate increases between 1980 and 1986, I use the Perl model to estimate penetration under the assumption that local rates remained at their 1980 levels. These predictions of telephone penetration are then compared with predicted values for 1987 at December 1986 rates.[8] Except for the very lowest-income category—an income of less than $10,000 (1986 dollars)—the predicted decline in telephone service from the higher real local rates is very low— often less than 2 percentage points even at incomes as low as $10,000 to $15,000 (1986 dollars). Additional analyses show that for married couples near the poverty level the effect of local rate increases between 1980 and 1986 is rather small, especially for older families. In short, the increase in the price of local access and usage in the early 1980s probably reduced telephone subscription at the lowest-income levels by about

7. Perl, "Residential Demand."

8. Real telephone rates peaked in 1986 and have fallen every year since then. Thus, the 1980–86 analysis should capture the effect of the period of rising real rates.

Table 5-1. Percent of Households with Telephone Service

Household income (1979 dollars)	Actual March 1987	Predicted: At December 1986 telephone rates	Predicted: At 1980 telephone rates
	Two married persons: white, urban		
Less than 5,000	89.0	91.2	92.7
5,000–9,999	94.0	95.4	96.2
10,000–14,999	97.2	97.0	97.6
15,000–19,999	98.5	97.8	98.2
20,000–29,999	98.7	98.6	98.8
30,000 or more	99.7	99.2	99.4
	Two married persons: black, urban		
Less than 5,000	79.0	80.0	83.0
5,000–9,999	91.8	89.1	90.9
10,000–14,999	92.6	93.1	94.4
15,000–19,999	97.8	94.5	95.5
20,000–29,999	95.3	96.3	96.9
30,000 or more	100.0	98.4	98.7
	Three-person household: white, urban		
Less than 5,000	70.2	77.3	80.7
5,000–9,999	84.7	86.6	88.9
10,000–14,999	90.8	92.5	93.8
15,000–19,999	96.5	95.7	96.4
20,000–29,999	98.6	97.9	98.3
30,000 or more	99.2	99.0	99.2
	Three-person household: black, urban		
Less than 5,000	90.2[a]	72.4	76.3
5,000–9,999	76.6	79.7	82.8
10,000–14,999	83.8	86.9	89.1
15,000–19,999	94.1	91.7	93.1
20,000–29,999	98.5	95.8	96.5
30,000 or more	97.7	98.3	98.2
All households	92.5	95.9	96.6

Sources: Bureau of the Census, Current Population Survey data base (March 1987); author's calculations based on data from National Association of Regulatory Utility Commissioners (NARUC), "Bell Operating Companies' Exchange Series Telephone Rates," Washington, 1986 (hereafter NARUC telephone rates); and Lewis J. Perl, *Residential Demand for Telephone Service, 1983* (White Plains, N.Y.: National Economic Research Associates, Inc., 1983).

a. Based on only twenty-three observations.

3 percentage points. The effect on lower middle-class households, however, is more modest.[9]

9. The Federal Communications Commission (FCC) and various states have developed low-income assistance programs to increase telephone subscription among lower-income households. Unlike policies that subsidize all residential access, these Lifeline and other programs benefit only lower-income subscribers.

These results are consistent with simulations conducted by Rolla Edward Park

One should not attribute all of the local rate increases between 1980 and 1986 either to divestiture or to FCC repricing decisions. Hence, even the 1 percent to 3 percent decline in telephone service for lower-income households may be an overestimate of the effects of the antitrust decree or the substitution of a subscriber-line charge (SLC) for interstate access charges assessed on interexchange calls through 1986

The Effect of Repricing on Income Distribution

As discussed, the shift in non-traffic-sensitive costs from interstate interexchange services to local service has raised the price of access to the network. In 1987 this repricing raised residential access charges by $2.46 billion and business access costs by $1.57 billion. In turn, the higher direct subscriber-line charges have permitted long-distance charges to fall by $1.63 billion for residential customers and by $2.44 billion for business customers. Thus, the average residential customer saw his or her bill rise by less than $10.00 in 1987. This increase for residential subscribers breaks down into an average 16.5 percent increase in the fixed monthly charge for access to the network and a 10.5 percent decline in interstate long-distance charges.[10]

The direct impact of repricing on residential consumers by income class in 1987 is summarized in table 5-2. All residential subscribers paid $27.60 a year in subscriber-line charges. These charges on residential and business consumers reduced average interstate long-distance charges by 10.5 percent. The savings in long-distance charges are estimated as follows. First, total telephone

and Bridger M. Mitchell, "Local Telephone Pricing and Universal Telephone Service," R–3724–NSF (Santa Monica, Calif.: Rand Corporation, June 1989). They find that a doubling of rates would reduce overall penetration by 5 percentage points and low-income penetration by 11 percentage points. Between 1980 and 1986, real local rates rose by about 30 percent; hence, my results should show an impact of about 30 percent of the Park-Mitchell rate-doubling exercise.

10. The average monthly charge for local service in 1987 is assumed to be $16.20 (FCC, Common Carrier Bureau, Industry Analysis Division, "Trends in Telephone Service," (Washington, August 1, 1988, p. 7). Without the $2.30 subscriber-line charge, this average would have been $13.90; therefore, the SLC raised local rates by an average of 16.5 percent. Total interstate long-distance charges in 1987 were $33.8 billion, assuming a full passthrough but ignoring elasticity effects. Thus, the $4 billion in SLC end-user charges reduced interstate long distance charges by about 10.5 percent.

Table 5-2. The Direct and Indirect Effects of Repricing of Telephone Service on Consumers by Household Income, 1987

Dollars per household per year[a]

Annual income	*Direct effect* *Local access*[b]	*Average long-distance use*[c]	*Indirect effect from other goods and services*	*Total effect*
Less than 5,000	27.60	−9.14	−2.77	15.69
5,000–9,999	27.60	−10.16	−2.64	14.80
10,000–14,999	27.60	−13.46	−3.57	10.57
15,000–19,999	27.60	−18.92	−4.33	4.35
20,000–29,999	27.60	−20.57	−5.79	1.24
30,000–39,999	27.60	−22.80	−6.91	−2.11
40,000 or more	27.60	−30.72	−11.82	−14.94

Sources: Author's calculations based on "Monitoring Report of the Federal-State Joint Board," in Federal Communications Commission CC Docket 87-339, January 1990, p. 246; Bureau of Labor Statistics (BLS), Consumer Expenditure Survey: Interview Survey, 1987 data base (August 1989), table 2; FCC, "News: Semi-Annual Study on Telephone Trends Brings Good News for Consumers," Washington, 1988; Bureau of Economic Analysis (BEA), "Input-Output Accounts BEA I D88-201: Industry by Commodity Total Requirements, 1982," Washington, 1988, table 5; and BLS, Consumer Expenditure Survey: Interview Survey, 1986 data base (February 1988), table 2. (Hereafter BLS Consumer Expenditure Survey 1986 data base.)

a. Assumes constant market basket for each income class.

b. Average subscriber-line charge of $2.30 a month times 12.

c. Assumes that local access costs average $16.20 a month and interstate interexchange equals 62 percent of total expenditures on telephone less local access costs.

outlays in 1987 are based on the BLS Consumer Expenditure Survey. Long-distance outlays are estimated by deducting the average annual local charges, $194.40, from the estimate of total telephone outlays. The interstate share of this long distance spending is assumed to be 62 percent. The total savings in long distance is equal to interstate long-distance outlays times 0.105. The results show that the annual savings for each household from reductions in interstate long-distance rates because of subscriber-line charges rise from $9.14 for the lowest-income category to nearly $31.00 for households with incomes of more than $40,000.

The indirect effects of telephone repricing derive from the reduction in business costs that are passed through to consumers.[11] By lowering the average cost of service to businesses, repricing not only provides more efficient signals to business consumers but it also lowers costs by 1.44 percent by 1987 and therefore reduces prices of telecommunications-intensive services. The consumer-services sectors of the economy with the greatest relative direct and

11. Long-distance rates fall sufficiently to offset the rise in business access costs. Obviously, if access costs rise sufficiently, some large business customers may choose to bypass the local network, thereby mitigating some of the higher local access costs.

Table 5-3. The Direct and Indirect Share of Purchased Telecommunications Services in Total Output

Sector	*Share of telecommunications services in output*
Finance, insurance	0.0357
Retail, wholesale trade	0.0236
Hotels, motels	0.0213
Health, education, social services	0.0182
Transportation	0.0160
Eating and drinking places	0.0147
Real estate, rentals	0.0053
All personal consumption expenditures	0.0180

Sources: Author's calculations based on BLS Consumer Expenditure Survey 1986 data base; table 2; and BEA, "Input-Output Accounts," table 5.

indirect consumption of telephone services are financial services; health, education, and social services; retail and wholesale services; transportation; hotels and motels; real estate and rentals; and eating and drinking establishments. Table 5-3 shows the total direct and indirect consumption of telephone services as a share of output for each of these sectors.

Consumer expenditures from the BLS Consumer Expenditure Survey were assigned to each of the input-output sectors listed in table 5-3. Nearly 75 percent of all consumer expenditures are estimated to pass through these sectors. Multiplying the direct and indirect telephone consumption in each dollar of output by these total expenditures results in an estimate of $403.25 a household in indirect consumption of telephone services through goods and services (other than telephone services) purchased in 1987, or 1.8 percent of average total consumer expenditures.[12] These indirect purchases of telephone service rise from 1.6 percent of income for most lower-income consumers to 1.9 percent for the highest-income class (over $40,000). Thus, the reduction in business telephone costs of 1.44 percent, if passed through fully to final product and services prices, reduces the average consumer's cost of other goods and services by about .023 percent. The resulting estimated savings by household income class in 1987, ranging from $2.77 to $11.82 a year, are shown in table 5-2.

12. This estimate for 1987 is based on the 1986 Consumer Expenditure Survey. Since the income distribution is recorded in nominal dollars, this one year's difference should have very little effect on the result.

The total effects of telephone repricing on income distribution, as it had evolved by 1987, are shown in the final column of table 5-2. As can be seen, the overall effect of telephone repricing on income distribution has been mildly regressive. The lowest-income classes paid about $16.00 more a year because of the repricing; the highest- income households saved about $15.00 a year.[13]

Thus, the overall effects on equity of repricing alone seem small. The lowest-income classes incur an increase in telephone-related expenditures of about 4 percent. The highest-income classes realize a savings of less than 3 percent. There is a redistribution from those using little interstate long-distance to those who use it intensively. But the shifts in the burden of telephone costs are so small— averaging less than 0.1 percent of income for each income class—that it is surprising that the subscriber-line charge was such a political issue.

Rise in Rural Rates since Divestiture

One problem purportedly created by competitive entry, the repricing of service, and the AT&T divestiture has been a rise in rural telephone rates. Rural access has been heavily subsidized by Rural Electrification Administration (REA) and Rural Telephone Bank loans as well as by the separations and settlements process of dividing interstate revenues. Because rural subscribers are probably more intensive users of interexchange services, the share of the non-traffic-sensitive subscriber plant costs allocated to interexchange calls was much higher for rural than for urban areas. For instance, in 1981, 56.5 percent of Wyoming's fixed subscriber plant costs were allocated to interstate services, but only 31.5 percent of New Jersey's fixed subscriber plant costs were allocated to the interstate jurisdiction.[14]

Obviously, the FCC's decision to reduce the subsidy to local service that flowed from excessive allocations of costs to interstate

13. These results are broadly consistent with the results that Lewis Perl obtained for cost-based pricing of all telephone service. See Lewis J. Perl, "Social Welfare and Distributional Consequences of Cost-Based Telephone Pricing," paper prepared for the Thirteenth Annual Telecommunications Policy Research Conference, Airlie, Va., April 23, 1985, p. 20.

14. Peyton L. Wynns, *The Changing Telephone Industry: Access Charges, Universal Service, and Local Rates* (Congressional Budget Office, 1984), pp. 12–13.

service could have an impact on rural telephone companies. By capping the subscriber plant factor used to allocate costs at 25 percent and shifting to a subscriber-line charge, the subsidy to high-cost rural exchanges was somewhat reduced. However, the FCC also provides "high-cost assistance" to local exchange companies with very high costs for each local line by allowing these companies to charge higher access fees to interexchange carriers.[15]

The effect of these policy changes is difficult to measure because data on rural rates are not collected in a systematic fashion. There are at least two ways to measure the changes in rural rates. One, examine the change in rates in the exchanges in the areas of the lowest population density.[16] These data are readily available for only the Bell operating companies and are tabulated by the National Association of Regulatory Commissioners. Two, try to measure the rate of change in local revenues for each subscriber for rural companies that borrow from the Rural Electrification Administration.

The Bell operating company data may be summarized for residential flat rates and the lowest monthly measured rate available in the smallest exchanges (rural) and the largest exchanges (urban) of each state (table 5-4).[17] By both measures rural rates seem to have increased much more rapidly than urban rates. The sample size for measured service is quite small because it was not available in many states in 1980, but these rates in 1988 were usually lower in real terms than flat rates in 1980. Thus, while local rates have risen, the real price of *access* by itself has not risen for those users who switched from flat-rate to measured service.

When one turns to the REA data, the pattern seems different. Between 1976 and 1988, the rate of increase in local rates seems much more modest. A rather straightforward regression analysis finds that average local revenue for each residential subscriber for a sample of one hundred REA-subsidized companies rose from

15. This policy, described in 47CFR Part 36, Subpart F, resulted from a Federal-State Joint Board recommendation to the FCC. (See FCC CC Docket 80-286.) See *Federal Register*, vol. 49, April 30, 1984.

16. This approach is suggested by Joseph F. Fuhr, "Telephone Subsidization in Rural Areas," Widener University, 1987.

17. The National Association of Regulatory Utility Commissioners provides data by size of local exchange in each state. The data in table 5-7 are for the smallest and largest exchange categories in each state. The lowest access rates are the rates offered for measured service.

Table 5-4. Local Rate Increases, 1980–88

Rate	Percent increase, 1980–88
Producer price index, local-residential rates	84.8
Consumer price index, local rates	94.1
Residential service, flat rate[a]	
Urban areas (largest exchanges; n = 49)[b]	68.6
Rural areas (smallest exchanges; n = 48)[b]	109.4
Residential service measured rate[c]	
Urban areas (largest exchanges; n = 21)[b]	42.2
Rural areas (smallest exchanges; n = 11)[b]	88.7
Overall consumer price index	43.6

Sources: Supplement to producer price indexes, annual issues; BLS consumer price index local rates reported in, "Monitoring Report of the Federal-State Joint Board," in FCC CC Docket 87-339, December 1988, p. 315, and January 1990, p. 235. Author's calculations based on NARUC telephone rates, 1980, 1988; and *Economic Report of the President, February 1990*, table C-58.

a. Weighted average of flat rates for Bell operating companies. Largest exchange and smallest exchange categories are for each of forty-eight states and the District of Columbia per NARUC data.

b. number of states for which data are available for 1980 and 1988.

c. Weighted average of measured (non-Lifeline) rate available for Bell operating companies. Data are for each of forty-eight states and the District of Columbia that had measured service in 1980 and in 1988.

$120 a year in 1976 to $138 a year in 1988.[18] Adding $31 for the subscriber-line charge and $6 for the implicit cost of a handset rental in 1988, the rate of increase in local costs for each subscriber is still less than half the average rate of inflation between these two years. Thus, the REA data show little evidence of an increase in residential rates for subsidized residential rural carriers. Apparently, rate increases for the smallest rural companies have not been as large as those for the smaller exchanges in the Bell companies.[19]

Maintaining Universal Service

Currently, universal service is not threatened by the decline in rural subsidies nor by the repricing of local and interexchange service. Approximately 93 percent of all households still have telephone service, and this share is growing.[20] Were local access prices lower, the share of households subscribing might be slightly

18. Data derived from Rural Electrification Administration, *Statistical Report, Rural Telephone Borrowers* (Department of Agriculture, 1976, 1988).

19. This is undoubtedly the result of the FCC's policy of using a "high-cost" formula in setting interstate access rates for small companies.

20. "Monitoring Report of the Federal-State Joint Board," in Federal Communications Commission CC Docket 87-339, July 1989, p. 3.

higher, but the effect of lower rates on even the poorest of households is very small.

Studies of low-income households have found that often the need to pay back bills, installation charges, and deposits dissuade lower-income consumers from subscribing or renewing their subscription to telephone service.[21] These charges are not related to the repricing of local and interstate service but rather to the fixed costs of connecting them to the network and assuring the carrier that they will pay their monthly bills.

To defray some of the costs of local access and installation charges, the Federal Communications Commission has developed two programs—Lifeline Assistance and Link-Up America.[22] Under the Lifeline Assistance program, begun in 1984 but enlarged in 1985, the FCC directs that some of the revenues from interstate access be used to defer the subscriber-line charges for low-income subscribers in a participating state if that state matches the amount of the SLC waiver. Under the Link-Up America program, begun in 1987, the FCC directs that the access pool provide up to $30 to defray up to one-half of the connection costs for eligible low-income households if the participating state matches the subsidy.

Perhaps it is too early to determine the effect of these programs. Twenty-six states and the District of Columbia had Lifeline Assistance programs by the end of 1988, and forty-five states and D.C. and Puerto Rico were participating in Link-Up America. The National Exchange Carriers Association estimated that these programs spent $76 million in total assistance in 1990.[23]

As Leland Johnson points out, it is not clear whether the Lifeline Assistance programs are designed to promote universal service or simply to provide rate relief to low-income consumers.[24] Moreover, the programs have relatively high administrative start-up costs. If the maximum value of Lifeline Assistance is double the current $3.50 subscriber-line charge, it would reduce local access costs by less than one-half. An elasticity of demand between −0.05 and

21. See, for example, the Michigan study cited in note 8.

22. For a description of those programs, see the "Monitoring Report" prepared by the staff of the Federal-State Joint Board in the FCC CC Docket 80-286.

23. "Monitoring Report of the Federal-State Joint Board," in Federal Communications Commission CC Docket 87-339, January 1990, pp. 56–58.

24. Leland L. Johnson, "Telephone Assistance Programs for Low Income Households: A Preliminary Assessment," R–3603–NSF/MF (Santa Monica, Calif.: Rand Corporation, February 1988).

−0.15 suggests a maximum impact of 7 percent of the targeted population. Even extended to the lowest quintile of the entire population, defraying twice the subscriber-line charge would only increase national telephone penetration by about 1.5 percent.

The modest set of programs of low-income assistance now in existence is surely preferable to the wholesale distortion of telephone rates to attempt to cross subsidize local access for a small share of the population. If direct subsidies can be targeted at low-income consumers, the inefficiencies in such a program will surely be less wasteful than the cross subsidization of all local service and the large subsidies to all rural consumers.

The Effect of Repricing on Business Users

The impacts of the recent repricing of telephone service have not been confined to residential consumers. The subscriber-line charge levied on business subscribers has been much greater than the residential subscriber-line charge. In addition, state regulators have attempted to mitigate the impacts of repricing on residential access rates by keeping intrastate toll rates from falling and by raising local business rates. Since businesses account for a large share of long-distance services, the impact of higher intrastate toll rates fall disproportionately on business.

The recent trends in local business rates provide an interesting case study in the politics of regulatory ratemaking. In 1980, for example, flat-rate business service was priced between 100 percent and 170 percent above flat-rate residential service (table 5-5). In 1988 this margin narrowed, but business rates were still far above residential rates. Recently, Bridger Mitchell has shown that the incremental capital costs of business service are somewhat lower than the costs of residential service in small to medium markets and perhaps 5 percent to 6 percent higher in the larger urban markets.[25] Since the variable noncapital costs are surely no higher for servicing and billing business customers, one must conclude that the enormous premiums for business customers are not driven by cost differences but by political considerations.

Equally important, Mitchell's results show that the smallest urban areas have capital costs per line that are nearly double those in the

25. Bridger M. Mitchell, "Incremental Capital Costs of Telephone Access and Local Use," prepared for the Incremental Cost Task Force, R-3764-ICTF (Santa Monica, Calif.: Rand Corporation, August 1989).

Table 5-5. Residential and Multiline Business Monthly Telephone Rates, 1980, 1988

Dollars per month per line unless otherwise noted

Type of exchange and rate	*December 1980*	*December 1988*	*Percent increase*
	Twenty-four most urbanized states[a]		
Largest exchanges			
Residential, flat rate (n = 25)[b]	9.87	16.34	65.6
Business, measured (n = 23)[b]	11.95	22.93	91.9
Business, flat rate (n = 19)[b]	25.13	30.25	20.4
Smallest exchanges			
Residential, flat rate (n = 24)[b]	6.63	13.19	98.9
Business, measured (n = 20)[b]	10.53	20.55	95.2
Business, flat rate (n = 21)[b]	13.32	28.30	112.5
	Twenty-four most rural states[a]		
Largest exchanges			
Residential, flat rate (n = 24)[b]	9.94	18.26	83.7
Business, measured (n = 12)[b]	17.23	30.45	76.7
Business, flat rate (n = 24)[b]	26.99	46.40	71.9
Smallest exchanges			
Residential, flat rate (n = 24)[b]	6.87	14.90	116.9
Business, measured (n = 12)[b]	12.01	24.26	102.0
Business, flat rate (n = 24)[b]	14.97	33.34	122.7

Sources: Author's calculations based on NARUC telephone rates, 1980, 1988.

a. Urbanized is measured by the share of a state's population in metropolitan areas. Excludes Alaska, Hawaii, and the District of Columbia.

b. Number of states for which data are available for 1980 and 1988.

larger urban areas. In fact smaller exchanges have consistently lower rates despite their higher costs (table 5-5).

Since 1980 the more rural states evidence a higher rate of increase in residential rates and business flat rates than the urbanized states, but the measured business rate increases are much more similar across the two groups of states. As with residential rates, business rates have risen more rapidly in the smallest exchanges than in the largest exchanges, narrowing the very large differentials that existed in 1980. Apparently, state regulators have been cognizant of bypass threats in the larger metropolitan areas and have therefore found it prudent to hold business rate increases down in these areas while forcing a large share of the burden of repricing on businesses in smaller communities.[26] Nevertheless, rates in the smallest exchanges

26. See Roger G. Noll and Susan R. Smart, The Political Economics of State Responses to Divestiture and Federal Deregulation in Telecommunications, Studies in Industry Economics, Discussion Paper 148 (Stanford University, May 1989).

remain lower than in the largest exchanges, suggesting that regulators still respond disproportionately to political pressures from more rural areas.

The Quality of Service

During the late 1960s, the quality of telephone service declined as local exchange carriers failed to keep up with the growth in demand. By the late 1970s, however, substantial new capital investment had restored the quality of service and even improved it.[27] When the break-up of AT&T was first announced, concern arose that the turmoil caused by the divestiture, the lack of coordination in systems planning, and consumer confusion would combine to lower the quality of telephone service. As of 1987, the evidence did not support such a concern.

Had local exchange carriers allowed the quality of service to decline, state public utility commissions would have received public pressure to investigate this degradation. A survey of the larger states' commissions revealed a lack of interest in the problem of quality. I called nine state commissions over a period of two years and found very little concern about quality.[28] Only Ohio admitted problems, and these problems concerned a non-Bell operating company. Maryland reported a few problems with quality in 1984, but none since then. New York collects information about the quality of service but does not process it into summary data. Virtually every state commission reported that the quality of service has not changed since 1984.

Data have also been collected from the seven regional Bell operating companies since 1985 by the Federal Communications Commission. These data provide detail on dial-tone delay, transmission quality, service order delay, equipment blockage and failure, and overall customer perception of quality. Most of the companies say they achieve between 90 percent and 99 percent of their technical service objectives in their operating jurisdictions. The only service-quality index showing a decline since 1985 is on-time service order

27. Andrew S. Carron and Paul W. MacAvoy, *The Decline of Service in the Regulated Industries* (Washington: American Enterprise Institute for Public Policy Research, 1981).

28. The state commissions surveyed were those for Illinois, Maryland, Pennsylvania, Florida, New York, Texas, Minnesota, Michigan, and Ohio.

completion for residential customers. The composite customer perception index for all companies rose from 91.5 percent to 94.0 percent between 1985 and 1988.[29]

The lack of quality degradation should come as no surprise. The surge in investment in fiber-optics capacity and equal access for interexchange services is conducive to an increase in quality. The new facilities offer improved signal quality, and more circuits are available to handle the increasing interexchange traffic.

In short, there is no evidence of problems with the quality of service since the first year of divestiture. Entry liberalization and divestiture have increased competition, and competitive firms respond to consumer demands for high-quality service. Liberalization has also increased consumer options and the complexity of monthly telephone bills, thus creating some confusion and increased search costs for consumers who had become accustomed to limited diversity and simple telephone bills.[30] However, some observers fear that the shift from rate-of-return regulation to price caps could alter the incentives for supplying quality in the next few years.[31]

The Effect of Regulation and Divestiture on Telephone Rates

The past two decades have seen great economic turmoil. Between the mid-1960s and 1980, inflation accelerated from 1.3 percent to 13.5 percent (table 5-6). This increase caused interest rates to rise steadily and then dramatically as the Federal Reserve attempted to stop the inflationary spiral in 1979. Tight monetary policy and the large budget deficits led to prolonged high real interest rates, recession, and a rising dollar during the early 1980s. These forces, in turn, led to sharp disinflation. By 1985 inflation rates had fallen to levels not achieved on a sustained basis since before the Vietnam War.

In an environment of rising inflation, regulators are often reluctant to allow rates to rise at the same pace as general inflation. This reluctance is reinforced by regulators' reliance on historical cost accounting conventions and the phenomenon of "regulatory lag." Although regulators must permit regulated carriers eventually to

29. These data are summarized in Jonathan M. Kraushaar, "Report on Quality of Service for the Bell Operating Companies," Federal Communications Commission, Common Carrier Bureau, Industry Analysis Division, Washington, March 17, 1989.

30. I am indebted to Leland Johnson for stressing this point to me.

31. See chap. 6 for a discussion of price caps.

Table 5-6. Market/Book Ratios for Telephone Company Equities and Inflation, 1964–88

	Market/Book Value						
Year	AT&T	GTE	Rochester Telephone	United Telephone	Continental Telephone	Standard and Poor's 400	Inflation[a]
1964	1.92	2.95	2.26	3.26	2.82	2.23	0.013
1965	1.63	3.34	2.33	3.45	3.52	2.26	0.016
1966	1.43	2.90	2.07	3.16	3.71	1.87	0.028
1967	1.25	2.54	1.84	3.26	3.47	2.20	0.031
1968	1.28	2.33	2.06	3.32	3.05	2.25	0.042
1969	1.14	1.77	1.56	2.15	1.98	1.96	0.054
1970	1.10	1.80	1.91	1.73	2.30	1.92	0.057
1971	0.98	1.72	1.87	1.70	2.04	2.04	0.044
1972	1.12	1.55	2.16	1.83	2.24	2.26	0.032
1973	1.02	1.23	1.33	1.29	1.45	1.74	0.062
1974	0.87	0.82	0.74	1.04	0.89	1.13	0.110
1975	0.94	1.17	0.80	1.12	1.01	1.42	0.091
1976	1.15	1.36	0.99	1.48	1.39	1.57	0.058
1977	1.01	1.24	1.00	1.36	1.14	1.27	0.065
1978	1.00	1.06	1.02	1.26	1.06	1.20	0.076
1979	0.82	0.99	0.98	1.26	1.12	1.23	0.113
1980	0.73	0.95	1.00	1.06	1.09	1.43	0.135
1981	0.87	1.07	1.28	1.21	1.12	1.18	0.103
1982	0.86	1.32	1.17	1.13	1.10	1.34	0.062
1983	0.88	1.29	1.25	1.09	1.28	1.52	0.032
1984	1.13[b]	1.12	1.28	1.17	1.27	1.51	0.043
1985	1.41[b]	1.43	1.43	1.35	1.34	1.86	0.036
1986	1.73[b]	1.68	1.60	1.45	1.49	2.16	0.018
1987	1.53[b]	1.48	1.48	1.62	1.58	2.13	0.036
1988	1.55[b]	1.79	1.76	2.54	2.03	2.17	0.041

Sources: Annual reports; Standard and Poor's data; and *Economic Report of the President, February 1990*, table C-58.

a. Rate of change in consumer price index, all urban consumers for all items.

b. For AT&T and divested regional holding (operating) companies.

recover their costs, this recovery need not occur immediately in response to inflationary pressures.

Andrew S. Carron and Paul W. MacAvoy have argued that during the inflationary 1960s and 1970s, regulators often squeezed regulated firms so much that they were forced to reduce service quality. Under their theory, as inflation recedes, real rates rise once again and quality recovers. For the telephone industry, however, the reductions in service quality—slower response to maintenance calls, slower dial tones, and increased circuit blockage—declined in the late 1960s but recovered during the inflationary 1970s. Carron and MacAvoy attribute this recovery to improved public-utility decisionmaking and increasing cross subsidies from long-distance to local service, but the change may not have been enough to offset fully the ravages of inflation on the regulated companies' balance sheets.[32]

If telephone regulators follow this pattern of suppressing rates during inflationary periods of declining inflation, one should expect to see telephone company equities falling in value relative to their book value as inflation accelerates and recovering during periods of relative price stability. Willard T. Carleton, Donald R. Chambers, and Josef Lakonishok have demonstrated that the ratio of market to book value for electric-utility equities fell noticeably in the 1970s as inflation accelerated and interest rates rose.[33] Telephone equities behaved similarly during this period.

The ratio of market to book value of equity for the principal telephone operating companies and holding companies fell dramatically between the mid-1960s and 1980. This decline was far more rapid than for the average industrial company, as one can see by comparing telephone equities with the S&P 400 index for (unregulated) industrial companies. The ratios of market to book value for AT&T and Rochester Telephone common stocks declined by more than 50 percent during this period, and the decline was even greater for United Telecom, Continental, and GTE. By contrast, this ratio declined by only 33 percent for the industrial firms in the S&P 400 between 1965 and 1980.

Since 1980 all telephone equities have recovered substantially (table 5-6) with the subsidence of inflation.

To determine whether recent rate increases have exceeded those that should be expected given the recent behavior of input prices and interest rates, a model of the process of telephone ratemaking is required. For this purpose, a public utility commission's regulation of telephone rates may be characterized in the standard fashion. The company's total return to capital ρ, must be allowed at least to equal the user cost of capital, r. This requires that:

$$\rho = (PQ - wL)/P_k K \geqq r \qquad (1)$$

32. Carron and MacAvoy, *Decline of Service*, pp. 43–44, 51.

33. Willard T. Carleton, Donald R. Chambers, and Josef Lakonishok, "Inflation Risk and Regulatory Lag," *Journal of Finance*, vol. 38 (May 1983), pp. 419–34.

where P is the average level of telephone rates, Q is the level of service, w is the cost of variable inputs, L is the level of variable inputs used in delivering Q, P_k is the cost of capital equipment, and K is the stock of capital used in delivering the service. Rearranging and dividing by Q:

$$(2) \qquad P \geqq r P_k(K/Q) + w(L/q)$$

The level of rates is determined by the cost of capital, r, the cost of capital equipment P_k, the amount of capital required per unit of output, K/Q, the level of variable input prices, w, and the amount of variable inputs (labor) per unit of output, L/Q. With technological progress, both K/Q and L/Q may decline over time. However, if there are increasing returns to scale, the ratios K/Q and L/Q may also decline with Q. Thus, variables capturing both the returns to scale and the pace of technological progress must be added to (2) to capture the determinants of rates over time.

To estimate the effect of input costs, output, and regulatory forces on the overall level of U.S. residential telephone rates, I relate telephone price levels to input prices, output levels, and inflation in the following manner:

$$(3) \qquad \begin{aligned} \text{Log}\, P_t = a_0 &+ a_1 \text{Log}\, Q_t + a_2 \text{Log}\, r_{t-1} + a_3 \text{Log}\, r_{t-2} \\ &+ a_4 \text{Log}\, P_{k,t-1} + a_5 \text{Log}\, P_{k,t-2} \\ &+ a_6 \text{Log}\, w_t + a_7 \text{TIME}_t + a_8 \text{INFLATION} \end{aligned}$$

where TIME is a time trend and INFLATION is the annual rate of increase in the consumer price index. This formulation is based on the assumption that regulators respond with a lag to a Cobb-Douglas approximation of a cost function.

Because P and Q are jointly determined, (3) must be estimated with a demand function:

$$(4) \qquad Q_t = f(P_t, \text{GNP}_t, \text{TIME}, \text{DIVEST})$$

where GNP is real gross national product and DIVEST is a dummy variable equal to unity for 1984-88 and zero for earlier years, capturing the effect of the detariffing of customer premises equipment (CPE).

To estimate (3), I use annual data for 1963–88. The dependent variable, P, is either a weighted average of real producer price indexes or the real consumer price index for telephone service.[34] The wage variable is the deflated value of average hourly earnings in the telephone services industry, SIC 481. The price of capital equipment (P_k) is the deflated value of the Bureau of Economic Analysis implicit price deflator for telecommunications equipment. The cost of capital (r) is approximated by the real bond rate, defined as the rate of return on newly issued telephone bonds minus a moving average of inflation rates.[35] Both r and P_k are lagged one year or two years.[36] The measure of real output, Q, is the deflated value of all telephone services from table 3-12.

The two-stage least squares estimates of (3) are reported in table 5-7. All results are corrected for serial correlation, using first-order corrections equal to ρ. Two dummy variables are included to test for the effects of divestiture—DIVEST and BREAKUP. The latter is equal to unity for 1982–88, the years since the *announcement* of divestiture.

The results of interest in table 5-7 are the estimated coefficients of INFLATION, DIVEST, and BREAKUP. In every case, the coefficient of INFLATION is negative and statistically significant, suggesting that regulators respond to inflation by suppressing real rates. In no case is the coefficient of DIVEST or BREAKUP statistically significant, but the magnitudes of each are greater for the consumer rate regression than for the producer-price regression. This result is at least consistent with the observed shift of regulatory costs from long- distance to local service.

In summary, it seems that much of the recent rise in real telephone rates can be explained by factor prices and the lagged effects of inflation. The increases in the early 1980s are largely a reflection of regulated firms catching up with past inflation. Nevertheless, this recovery in rates has not been sufficient to allow market/book ratios for most telephone equities to recover to their

34. The weights are the 1988 revenue shares for each of the six categories of producer price indexes (PPIs)—interstate MTS, intrastate MTS, interstate WATS, intrastate WATS, local residential, and local business. These PPIs are only available for the years since 1972.

35. Specifically, r is equal to the telephone bond rate less a weighted average of current inflation and the last three years' inflation rates with weights of 0.4, 0.3, 0.2, and 0.1 on the current and successively lagged inflation rates. Both r and P_k are lagged one and two years to reflect the adjustment of telephone rates to the real rental rate of capital.

36. The results in table 5-7 are reported for only one-year lags on P_k and r for the producer price regressions because of the limited degrees of freedom available.

Table 5-7. Two-Stage Least-Squares Estimates of the Determinants of Telephone Price Levels[a]

Independent variable	*Dependent variable*				
	Log real producer prices		*Log real consumer prices*		
	(1)	*(2)*	*(3)*	*(4)*	*(5)*
Constant	3.43	3.49	5.63	5.31	5.54
Log Q	−0.48	−0.46	−0.28	−0.24	−0.24
	(7.28)	(7.04)	(4.49)	(4.30)	(4.66)
Log r_{-1}	0.015	0.021	0.013	−0.0010	0.014
	(1.53)	(1.67)	(1.02)	(0.05)	(1.03)
Log r_{-2}	...	...	0.023	0.038	0.037
			(1.82)	(3.01)	(3.06)
Log $P_{k,-1}$	0.20	0.24	0.35	0.32	0.40
	(1.45)	(1.98)	(1.89)	(1.54)	(2.03)
Log $P_{k,-2}$	...	...	0.28	0.50	0.39
			(1.43)	(2.14)	(2.01)
Log w	0.28	0.33	0.010	−0.033	0.05
	(1.35)	(1.47)	(0.05)	(0.17)	(0.30)
INFLATION	−0.41	−0.44	−0.79	−0.79	−0.86
	(2.25)	(2.03)	(2.85)	(2.69)	(3.70)
DIVEST	0.014	...	0.031	...	...
	(0.83)		(1.42)		
BREAKUP	...	−0.0080	...	0.040	...
		(0.29)		(0.92)	
Summary statistics:					
$\bar{R}^2$	0.990	0.989	0.993	0.993	0.993
DW	1.616	1.537	1.602	1.804	1.745
ρ	0.201	0.282	0.550	0.344	0.273
Period	1972–88	1972–88	1963–88	1963–88	1963–88

a. The numbers in parentheses are *t*-statistics.

levels in the mid-1960s. The loss of CPE markets, the obsolescence of capital equipment, and increasing competition may preclude a return to the mid-1960s relative values of these equities.

Divestiture, Competition, and the Systematic Risk of Holding Telephone Equities

Public utility regulation usually offers regulated firms some protection from the threat of entry in return for limiting their pricing discretion. Throughout the 1960s and 1970s, AT&T and other telephone carriers found themselves in a changing regulatory environment in which the protection against the threat of entry was no longer assured.

One of the costs of liberalization of entry and the pursuit of the U.S. antitrust suit was undoubtedly the increase in risk premiums demanded by investors to extend capital funds to the telephone sector. In this section, I examine the changes in risk confronting owners of telephone equities from 1955 to the present. For this purpose, I utilize the capital-asset pricing model's estimates of the systematic risk of holding equities, β.[37]

Although the new competitors to AT&T were not quantitatively important before divestiture, the mere threat of entry was apparently unsettling to investors in AT&T stock. Surprisingly, the systematic risk from holding AT&T common stock rose considerably between 1959 and 1962 after the FCC first admitted private microwave operations in the Above-890 proceeding (table 5-8). The estimate of β rose from 0.29 in the 1955-58 period to 0.96 in the 1961-62 period.

After 1962, as the private microwave threat appeared to subside, AT&T's β fell to 0.53. However, when the MCI entry into private-line common carriage was allowed in 1969 and the commission ruled in favor of Carterfone, the systematic risk of holding AT&T common rose once again to 0.89. Clearly, investors viewed AT&T as a much more risky investment during the 1969–74 period as trustbusters and regulators began to act as if interstate telecommunications should be a competitive market.[38]

A closer look at the 1959–83 period is even more illuminating. Table 5-8 shows the market's estimate of the systematic risk from holding AT&T common and the excess returns realized from this equity in successive two-year periods. Once again, the response to perceived entry threats is clearly enormous. In 1959–60 and 1969–70, β more than doubled from the respective preceding period. The excess returns, however, show a somewhat different pattern. Between

37. The coefficient, β, is the estimate of the covariance of any risk asset's return with the overall market portfolio of risky assets divided by the variance in the market portfolio.

38. This rise occurred even though the FCC was denying that it would allow competition in switched interstate services. Apparently, the equity market did not fully accept these assurances.

Table 5-8. Risk and Excess Return for Investments in AT&T Common Stock before Divestiture

Period	Systematic risk (β)	Average monthly excess returns using predictions from: 1955–58 CAPM EQ.	1955–62 CAPM EQ.	Average long-term corporate bond rate (percent)
1955–58	0.29	...	...	3.52
1959–60	0.70	+0.009	...	4.40
1961–62	0.96	−0.002	...	4.34
1963–64	0.53	−0.003	−0.006	4.33
1965–66	0.63	−0.015	−0.013	4.81
1967–68	0.44	−0.010	−0.013	5.84
1969–70	0.89	−0.007	−0.005	7.54
1971–72	0.81	−0.003	−0.006	7.30
1973–74	0.61	−0.006	+0.002	8.00
1975–76	0.57	+0.005	0.000	8.63
1977–78	0.36	−0.006	−0.004	8.38
1979–80	0.15	−0.002	−0.002	10.78
1981	[a]	+0.012	+0.017	14.17
1982–83	[a]	−0.006	−0.008	12.92

Sources: Author's calculations based on the University of Chicago, Graduate School of Business, Center for Research in Security Prices, New York Stock Exchange-American Stock Exchange (NYSE-AMEX) files data base (various years); Ibbotson Associates, *Stocks, Bonds, Bills, and Inflation: Market Results for 1926–1989* (Chicago, 1990); and *Economic Report of the President, February 1990*, table C-71.
a. Not statistically significant.

the end of 1964 and the end of 1970, the value of AT&T stock fell by 31 percent relative to predictions from the 1955–62 estimates of the capital asset pricing model.

Part of the 1965–70 decline in the relative value of AT&T may be attributed to rising interest rates. Between 1964 and 1970, the long-term bond rate rose from 4.4 to 8.0 percent as inflation accelerated. But inflation cannot be the principal explanation for the rising β, given that β fell for AT&T between 1970 and 1974, a period of great acceleration in inflation.

Further insights may be gleaned from estimates of β for several different telephone companies, including two other holding companies, GTE and Continental and two operating companies, Rochester and New England Telephone (table 5-9). GTE and Continental are viewed by the market as riskier than AT&T throughout the period despite the absence of major antitrust actions against the former companies. Telephone operating companies appear to be less risky than the more diversified companies such as GTE and AT&T. But

Table 5-9. Estimates of the Systematic Risk (β) of Holding Telephone Equities, 1955–88

Period	*AT&T*	*Regional Bell operating companies*	*GTE*	*Contel*	*Rochester*	*New England T&T*
1955–58 (Before private microwave)	0.29	...	0.77	...	...	...
1959–62 (Before MCI applications to FCC)	0.87	...	1.17	...	0.85[a]	...
1963–68 (Before MCI entry authorization and *Carterfone*)	0.54	...	0.96	...	0.63	0.44
1969–1974(06) (Before antitrust suit, Execunet)	0.78	...	1.44	1.38	0.82	0.58
1974(07)–1981 (Before antitrust settlement)	0.36	...	0.60	0.82	0.41	0.48[b]
1984(03)–88	0.68	0.53–0.71	0.68	0.90	0.42	n.a.

Sources: Author's calculations based on data from the University of Chicago, Graduate School of Business, Center for Research in Security Prices, NYSE-AMEX files data base (various years). (Hereafter University of Chicago, Security Prices data base.)

n.a. Not available.

a. 1959(04)–1962.

b. 1974(07)–1980.

all telephone-related stocks appear to suffer increases in β during periods of regulatory upheaval. Between 1974 and 1981, β's fell for all telco stocks, but since 1981, β's have risen once again, suggesting that investors view the current climate of telephone regulation with increased wariness.

The divested regional Bell operating companies' common equities have a range of estimated β that vary from 0.50 to 0.71, but no clear relationship exists between cumulative returns since divestiture and β (table 5-10). Despite β's of less than one, AT&T's common stock and every regional Bell operating company equity has outperformed the overall market since divestiture.

Finally, AT&T now has a higher β and lower total returns than the average regional Bell company. The equity markets thus appear to be saying that AT&T is both slightly more risky than the regional

Table 5-10. Performance of the Divested Regional Bell Holding Companies and AT&T, 1984(03)–1988(12)

Company	β	*Cumulative return*
Ameritech	0.53	193.0
Bell Atlantic	0.55	179.2
Bell South	0.71	158.8
Nynex	0.66	200.7
Pacific Telesis	0.62	211.7
Southwestern Bell	0.61	181.4
U.S. West	0.68	172.4
AT&T	0.68	120.2
All stocks	1.00	113.6

Sources: Author's calculations based on data from the University of Chicago, Security Prices data base; and Ibbotson Associates, *Stocks, Bonds, Bills, and Inflation.*

Bell operating companies and less profitable relative to expectations in early 1984.[39]

Divestiture and the Changes in Industry Output

If divestiture had caused telephone rates to rise precipitously or service quality to fall, one might expect the quantity of service to decline. Thus, one might expect some decline in output relative to the output that would have obtained without the divestiture if the break-up of AT&T raised rates and lowered the quality of service. Moreover, if the divestiture sacrificed economies of scale and scope or otherwise created inefficiencies, a decline in productivity in delivering telephone services should have occurred. As discussed, there is little evidence that divestiture has been responsible for deterioration in service quality or productivity. Possibly, however, alternatives to common carriage have become increasingly attractive in recent years. If so, a decline in the output of total regulated carriers may have occurred, compared with what one might have expected given the level of rates and economic activity.

What has happened to total output of telephone services since divestiture? To analyze this question, I estimate the reduced form of equations (3) and (4), using time series data for 1963–80, and I

39. In 1989, however, AT&T outperformed the equities of most regional Bell operating companies.

compare 1981–88 predictions with actual output. If customers are shunning the network in favor of private carriage, the reduced form equation should overpredict 1981–87 output levels. The differences in the logarithms of actual and predicted output are as follows:

1981	−0.008
1982	−0.087
1983	−0.114
1984	−0.197
1985	−0.230
1986	−0.220
1987	−0.212
1988	−0.248

These shortfalls are very close to my earlier estimates of the share of investment occurring outside the network.[40] The deviation from predicted output rises sharply in 1984, the year of detariffing the Bell System's CPE, and it continues to rise through 1988 to nearly 25 percent. This result suggests that an increasing share of telecommunications activity is taking place in private business networks.

The Impact on Economic Welfare

In this section I estimate the effects of liberalization and divestiture on overall economic welfare. Because many of the policy changes are still evolving and because empirical estimates of several changes are difficult to obtain, the estimated welfare gains should be viewed as only a part of the potential gains from liberalization.

The Sources of Welfare Gains and Losses

There are at least seven sources of potential welfare improvements from entry liberalization and divestiture and the regulatory decisions precipitated by these events: improvements in productive efficiency in the telecommunications services sector; lower telecommunications equipment prices; the repricing of access and long-distance services; increased use of local measured-service pricing; the reduction in the discrimination against business in access pricing; the reduction in the discrimination in favor of rural areas; and the increase in the variety of new services and equipment caused by competitive entry.

40. See table 3-3.

Not all of these changes, however, are likely to have contributed significantly toward improvements in welfare as yet. The reduction in discrimination against business and urban subscribers has been very slow to materialize. Moreover, measuring the impacts of new equipment and new services is difficult because their values are difficult to observe and the "but-for" scenario is extremely difficult to model. As a result, most of my estimated improvements in economic welfare derive from the improvements in productive efficiency and the repricing of local access and toll services.

IMPROVED PRODUCTIVE EFFICIENCY. The introduction of competition into interstate telephone service and terminal equipment markets induced big improvements in efficiency after 1971. As discussed in chapter 3, productivity growth was relatively constant in the telecommunications sector in the 1960s, but total factor productivity (TFP) began to accelerate after 1971, just as entry arrived in long-distance services and three years after the *Carterfone* decision induced at least limited competition in terminal equipment.

It would be difficult to ascribe all of the acceleration in TFP to these first moves toward liberalization, but it is also difficult to find any other explanation for the acceleration. Microwave developed in the 1950s, not in the 1970s. The electronics-computer revolution was certainly in full flower during the 1960s. The 1970s were not a period of generally rapid technical change in the U.S. economy.

Because I cannot establish with precision the exact contribution of liberalization to technical change and TFP growth in the telecommunications sector, I simply estimate the savings in total costs for the telecommunications sector from 1972 to 1988 from the estimates of the acceleration in productivity shown in table 5-11.[41] The effect of the acceleration in productivity growth obviously grows over time. By 1988 total annual telecommunications sector costs are $3.5 billion lower, owing to the greater improvement in technical efficiency. Thus, even if only half of this improvement can be attributed to the competitive spur of liberalization, the annual net benefit by 1988 is $1.75 billion.

Divestiture itself offset the general improvement in efficiency in 1984 and 1985, apparently because of the wrenching adjustments it caused within the Bell companies. This offset totaled $3.7 billion,

41. These calculations are based on estimated technical progress and output indexes from chap. 3.

Table 5-11. Estimated Cost Saving from Technological Progress, Competition, and Divestiture, 1972–88

Year	*Output index (1988 = 1)*	*Proportional saving because of TFP growth*[a]	*Estimated cost saving from TFP growth (billions of 1988 dollars)*[b]	*Losses in 1984–85 because of divestiture (billions of 1988 dollars)*[a]
1972	0.356	0.0019	0.073	...
1973	0.391	0.0038	0.161	...
1974	0.420	0.0057	0.259	...
1975	0.440	0.0076	0.361	...
1976	0.468	0.0095	0.480	...
1977	0.514	0.0115	0.638	...
1978	0.570	0.0134	0.825	...
1979	0.633	0.0153	1.046	...
1980	0.693	0.0172	1.287	...
1981	0.732	0.0192	1.518	...
1982	0.753	0.0211	1.716	...
1983	0.770	0.0230	1.913	...
1984	0.767	0.0250	2.071	−1.786
1985	0.804	0.0269	2.336	−1.872
1986	0.863	0.0289	2.694	...
1987	0.945	0.0308	3.143	...
1988	1.000	0.0327	3.532	...

Source: Author's calculations based on table 3-12, industry data, and equation 2, chap. 3.
a. From equation 2, chap. 3.
b. The estimated cost saving is $108 billion times output index times the proportional saving.

or about one year's current estimated improvement in the industry's productive efficiency. Thus, over the 1980s (through 1988) the maximum net improvement in technical efficiency that could be ascribed to liberalization was $16.6 billion. If only half of the TFP acceleration were attributed to liberalization, this total would fall to $6.4 billion.[42]

LOWER EQUIPMENT PRICES. The data on telecommunications equipment prices are unfortunately too flawed to permit an accurate assessment of the effects of liberalization. Before 1985, most government estimates of telco equipment were based on the Bell

42. A recent paper by John S. Ying and Richard J. Shin, "Costly Gains to Breaking Up: LECs and the Baby Bells," University of Delaware, Department of Economics, October 1989, uses a translog cost model to estimate the effects of divestiture on the efficiency of local carriers. They find that cost savings of nearly 4 percent were realized by 1987, but this result is likely to be biased upward because the authors apparently fail to adjust for detariffing of customer premises equipment (CPE) and inside wiring and AT&T's write-off of substantial capital stock at the end of 1983.

telephone plant index (TPI), which failed to consider quality improvements between generations of equipment. Recent government price series seem to be at variance with market research estimates.

The absence of a separate estimate of the welfare effects of lower equipment prices may not be a serious problem, given the estimate of efficiency improvements in telecommunications services. As mentioned in chapter 3, the estimates of TFP increases in the equipment-using sector may be biased upward because of an upward bias in the telecommunications equipment deflator over time. Of course, this offsetting effect would not carry over into customer-owned customer premises equipment (CPE). Thus, the estimates of welfare gains in the improvement of production efficiency may include substantial benefits of lower telco equipment prices. It is irrelevant for economic welfare whether the improvements in efficiency derive from lower equipment prices or from greater efficiency in using this equipment.

THE REPRICING OF TOLL AND LOCAL SERVICES. The FCC's response to the growing distortions in pricing long-distance service began before divestiture but after the other common carriers (OCCs) began to offer a serious competitive threat to AT&T. By moving rates toward the costs of individual services, the FCC has achieved a great improvement in economic welfare.

The static welfare effects of repricing are registered in three distinct markets: residential access, business access, and interstate toll. The changes in economic welfare in each of these three markets depend on the magnitude of the price change, the elasticity of demand, and the incremental cost of service. While reasonably satisfactory estimates of the first two of these magnitudes for each market exist, the data on incremental costs are surprisingly scarce in this regulated industry.

The rudiments of the repricing welfare analysis are shown in figure 5-1. The prices of the three services are given by P^{RA}, P^{BA}, and P^{IT}—for residential access, business access, and interstate toll. The subscripts *b*, *a*, and *f* are used to represent the prices before the FCC's subscriber-line charge, after the imposition of the SLC, and with the full imposition of an SLC that is sufficient to cover all federally assigned non-traffic-sensitive costs. These three alternatives allow one to compare the welfare effects of the actual FCC policy with the originally intended policy of total repricing of interstate toll.

Figure 5-1. Welfare Effects from Repricing Telephone Service

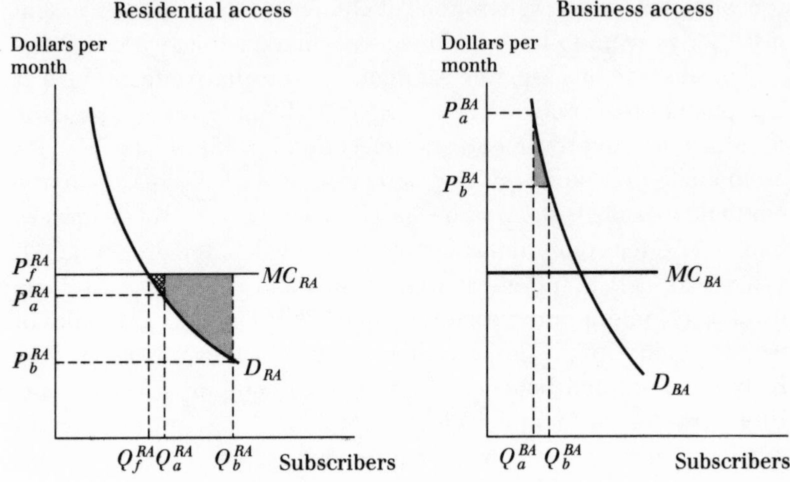

Interstate interexchange market

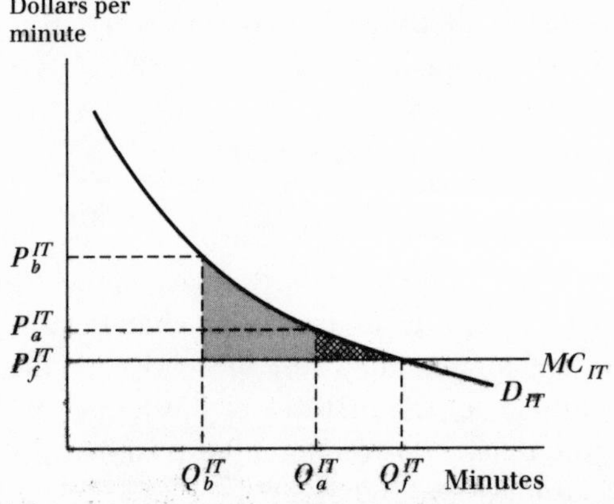

In each panel of figure 5-1, the static welfare effects of actual FCC policy and full repricing are shown by the shaded and cross-hatched areas, respectively. For example, the rise in local residential access rates reduces total residential subscriptions from Q_b^{RA} to Q_a^{RA}. If the incremental cost of residential access, MC^{RA}, is above the price of service, this increase in rates caused by the SLC improves economic welfare by an amount equal to the shaded area between the demand curve, D^{RA}, and incremental costs, MC^{RA}.

Similarly, the reduction in the interstate toll rate from P_b^{IT} to P_a^{IT}, caused by the reduction in carrier access charges, increases total interstate usage from Q_b^{IT} to Q_a^{IT}. The resulting increase in consumer welfare is equal to the area between the demand curve and the incremental cost locus, MC_{IT}, evaluated between Q_b^{IT} and Q_a^{IT}. The sum of the welfare effects across all three markets reflects the net gain (or loss) from repricing.

There are at least two complications to the simplified welfare analysis depicted in figure 5-1. First, the demands for the various types of service are not independent. The demand for toll services depends on the number of subscribers on the system and, therefore, on the price charged to business and residential subscribers for access to the network. Similarly, the demand for access is likely to be a function of the price of usage—including interstate toll usage. These interdependencies could be easily accounted for in figure 5-1 by shifting the demand curves for access outward as the price of toll service is lowered and shifting the demand curve for toll service inward to reflect the effect of higher subscriber access prices. Unfortunately, the lack of good estimates of these cross-price effects makes it impossible to include such effects in our empirical analysis. This omission is not likely to be very important, however, because the subscribers who place a low enough value on telephone service to be excluded by the rising price of access are unlikely to generate very much toll activity and are also unlikely to be heavily affected by lower toll rates in deciding to subscribe.[43]

In addition, the decision to subscribe to the telephone network creates consumption externalities since the value of telephone service to others is enhanced by another subscriber to the network. This externality can easily be accounted for by reducing the incremental cost of service by an estimate of the value of the incremental subscriber to others.[44]

43. See Jeffrey Rohlfs, "Economically-Efficient Bell-System Pricing," Bell Laboratories, Economic Discussion Paper 138, January 1979, for an estimate of these cross-price effects. He finds that rather rough estimates of the cross-elasticities increase the "superelasticities" for local and switched-message telephone services somewhat, but reduce the superelasticity for WATS.

44. For another approach to this problem, see Lewis J. Perl, "Social Welfare and Distributional Consequences of Cost-Based Telephone Pricing," paper prepared for the Thirteenth Annual Telecommunications Policy Research Conference, Airlie, Va., April 23, 1985.

Table 5-12. Parameter Values for Repricing Analysis

Item	Value
Residential access	
Number of lines	90,184,000
Average annual price (with 1988 SLC)	$198.84
Price elasticity of demand	−0.1
Incremental cost per year	$150 or $240
Value of externality per year	$48
Business access	
Number of lines	39,525,000
Average annual price (with 1988 SLC)	$481.44
Price elasticity of demand	−0.05
Incremental cost per year	$150 or $240
Value of externality per year	$48
Interstate toll	
Number of annual conversation minutes	180.1 billion
Price per minute	$0.167
Price elasticity of demand	−0.7
Incremental cost per minute	$0.08 or $0.137

Sources: See text.

To estimate the welfare effects of repricing, estimates of a number of parameters must be drawn from data reported by the FCC or from the empirical literature (table 5-12). For cost data, I rely on reported accounting estimates of average costs or engineering analyses of incremental costs. The best estimate of the *capital* costs of local access is from the recent work of Bridger Mitchell based on an engineering analysis of California carriers.[45] He finds that the incremental capital costs of residential service fall between $102 and $113 per year in smaller urban areas to as little as $53 to $72 for larger cities. The incremental cost of business lines is slightly lower in all but the larger cities. Unfortunately, Mitchell's study does not include marginal operating costs, but an econometric analysis by Rohlfs of one urban telephone system estimated these operating costs to be about $130 a year (1983 dollars).[46] If Rohlfs's and Mitchell's studies are combined, the incremental cost of access and local usage seems to be between $200 and $240 per year, depending on the density of the market.

45. Mitchell, "Incremental Capital Costs."

46. Jeffrey H. Rohlfs, "Marginal Costs of Telephone Services in Washington, D.C." Shooshan and Jackson, Inc., Washington, October 1983. It should be noted that Rohlfs forecast that the marginal capital costs of residential access in the District of Columbia in 1984 were $470 a year (p. 35).

A study of Indiana Bell found much lower incremental costs of access and local usage than those reported by Mitchell and Rohlfs.[47] It concluded that the long-run marginal cost of service varied from $18.00 to $81.00 per year for residential customers and from $8.40 to $83.00 for business customers. The reported average, unseparated, non-traffic-sensitive costs for all local exchange carriers, however, averaged $230.54 in 1988.[48] Thus, one is left with a wide range of estimates of the cost of local service. Given Mitchell's recent results, it seems unlikely that the incremental cost of telephone access and local use including incremental operating costs could be much below $150 in all but the largest markets. The average accounting costs of access, however, are reported to be $230.54. Since the traffic-sensitive costs of providing access and local usage are very low, $240 is used as the upward bound for the incremental cost of residential and business access and local usage.[49]

The demand elasticities are drawn from Lester Taylor and Lewis J. Perl. All demand functions are linear in the logarithms. The estimate of the demand externality is based on work by Perl.[50] All estimates of interstate toll usage are based on FCC reports of subscriber-line usage for interstate services. The average price for each minute for interstate toll is obtained by adjusting Wenders and Egan's average AT&T interstate rate for 1983 by the change in the producer price index for interstate message telecommunications service (MTS).[51] The average price for each minute for interstate WATS is then calculated from estimated subscriber line usage and reported revenues for interstate toll services. The price of interstate toll is then calculated as a weighted average of WATS and MTS. The incremental cost for each minute of interstate toll is either the value reported by Wenders and Egan or the current price less the

47. Ben Johnson Associates, Inc., "The Marginal Cost of Local Telephone Service in Indiana," Tallahassee, Fl., November 1986.

48. "Monitoring Report of the Federal-State Joint Board," January 1990, p.87.

49. Park and Mitchell, "Local Telephone Pricing." In his 1985 study, "Social Welfare," Perl used $261 and $277 for the cost of access for residential and business customers, respectively.

50. Taylor, *Telecommunications Demand;* and Perl, "Social Welfare." See also the estimates reported by Rohlfs, "Marginal Costs"; and John T. Wenders and Bruce L. Egan, "The Implications of Economic Efficiency for U.S. Telecommunications Policy," *Telecommunications Policy* vol. 10 (March 1986), pp. 33–40.

51. Wenders and Egan, "Implications of Economic Efficiency."

remaining non-traffic-sensitive costs.[52] Local rates are based on FCC estimates, as reported in the January 1990 *Monitoring Report.* The average business rate is for single-line customers only because there are no data on multiline customers.[53] The number of residential and business subscribers is obtained from the FCC's *Statistics of Communications Common Carriers* and the U.S. Telephone Association's *Statistics of Local Exchange Carriers, 1989.* All nonresidential lines are assigned to "business."

The estimated welfare gains from repricing are shown in table 5-13. Under the most optimistic assumptions about costs the net welfare gains for the 1988 repricing are more than $1.4 billion a year. A full shift of non-traffic-sensitive costs from long-distance to local access would more than double these gains to $3.1 billion a year. If incremental costs approximate current accounting costs, the welfare gains are more modest, ranging from $0.7 billion to $1.0 billion a year. In either case, the benefits from repricing just interstate service are substantial; the benefits of repricing intrastate toll would be even greater. The welfare gains come almost entirely from long-distance services, and they do not vary much with changes in the assumptions about the marginal cost of local access.

If one uses Wenders and Egan's estimates of the marginal cost of intrastate and interstate toll and assumes that the marginal cost of additional access lines is $150 a year, one may estimate the gains from "quasi-optimal" pricing of residential access and toll.[54] Toll

52. The latter estimate assumes that the current interstate toll market is fully competitive and that incremental costs are equal to currently realized average costs. Perl's 1985 study used a range of costs of 7 cents to 9 cents a minute and 13 cents to 14 cents a minute, very close to the nominal costs used in this book.

53. This is not a serious problem because the large, multiline businesses often obtain discounts on Centrex services that offset the higher multiline subscriber-line charges. Multiplying the average residential and single-line business rates by the total number of residential and business lines, respectively, yields approximately $37 billion for 1988 compared with $40 billion in local revenues reported by the industry. Author's calculation based on FCC, *Statistics of Communications Common Carriers* (Washington, 1988-89), tables 2-5, 2-9; and U.S. Telephone Association, *Statistics of the Local Exchange Carriers* (Washington, 1989), pp. 2, 11. Information on average residential and single-line business rates from FCC, Common Carrier Bureau, Industry Analysis Division, "Trends in Telephone Service," Washington, August 1990, pp. 8–9.

54. These estimates are $0.05 a minute for intrastate toll and $0.08 a minute for interstate toll. "Quasi-optimal" refers to Ramsey pricing under a revenue constraint. See William J. Baumol and David F. Bradford, "Optimal Departures from Marginal Cost Pricing," *American Economic Review*, vol. 60 (June 1970), pp. 265–83. I ignore any effects on welfare from changes in local usage.

Table 5-13. Estimated Welfare Gains from Repricing Interstate Toll and Access, 1988

Billions of 1988 dollars

Item	*1988 subscriber-line charge*	*Full subscriber-line charge*
Using lower bound for incremental costs		
Residential access	−0.128	−0.316
Business access	−0.035	−0.097
Interstate toll	1.597	3.504
Total	1.434	3.091
Using upper bound for incremental costs		
Residential access	0.016	−0.027
Business access	−0.026	−0.075
Interstate toll	0.674	1.053
Total	0.664	0.951

Source: Author's calculations based on table 5-12.

rates would fall to $0.089 and $0.0556 a minute for interstate and intrastate toll, respectively, while the annual charge for residential access would rise to $355. The result would be another $10.9 billion in welfare gains over and above the $1.4 billion already gained from the 1988 subscriber-line charge. This figure is very close to Wenders and Egan's estimates of $10.7 billion for 1983 and Perl's "high-elasticity" estimate of $10.8 billion for 1985.[55]

LOCAL MEASURED-SERVICE PRICING. The gradual movement toward local measured-service pricing occurred at least partially because of liberalization. When the new competitive long-distance carriers first began to use the Bell and independent circuits to terminate their calls, they often used local circuits that were priced on a flat-rate basis. Later, the FCC negotiated a special ENFIA tariff for the inferior local connections of these new OCCs, thereby reducing the revenues available to local carriers from long-distance services. More recently, the pressure deriving from the FCC's use of subscriber-line charges to pay for non-traffic-sensitive costs of local circuits has led state regulators to allow limited introduction of local measured service in some jurisdictions.

The spread of measured service has been impeded by regulatory restraints on pricing the various options. Unless the flat-rate service

55. The $12.3 billion total in 1988 dollars is equal to $11.2 billion in 1985 dollars or $10.4 billion in 1983 dollars using the consumer price index as a deflator.

carries a big premium over the minimum monthly rate for measured service, subscribers will not be induced to choose measured service. In many jurisdictions, measured service is not offered to residential subscribers at all. In still others, it is offered only at a mild discount from flat-rate service and thus is not chosen by many subscribers. Consequently, only 11 percent of the nation's residential subscribers had measured service in 1988.[56]

James M. Griffin and Thomas H. Mayor have estimated that local measured service can yield as much as $800 million in annual welfare improvements, but their estimate falls rapidly as the share of subscribers opting for measured service declines.[57] As a practical matter, so few subscribers are now taking measured service that the total welfare gains from this option are de minimis. Moreover, if incremental capacity cost is low and monitoring costs are substantial, the benefits from measured service pricing may be very modest or even nonexistent.[58]

REDUCTION IN THE DISTORTION BETWEEN RESIDENTIAL AND BUSINESS RATES. Historically, regulators have allowed telephone companies to charge business subscribers higher rates per line than residential subscribers in the same area. Recent research has shown that the fixed (non-traffic-sensitive) cost of serving business customers may even be lower than the cost of offering residential service in the same locality.[59] Thus, the traditional pricing pattern can only be justified as a form of second-best, inverse-demand-elasticity pricing, assuming that business customers have a lower price elasticity of demand than residences.

With the development of privately owned CPE and various bypass

56. NARUC data for December 1988.

57. See James M. Griffin and Thomas H. Mayor, "The Welfare Gain from Efficient Pricing of Local Telephone Services," *Journal of Law and Economics*, vol. 30 (October 1987), p. 482. A more recent study by Park and Mitchell, however, finds that the welfare gains from measured-service pricing are more than offset by monitoring costs at current flat-rate service prices. (Park and Mitchell, "Local Telephone Pricing.")

58. Park and Mitchell, "Local Telephone Pricing." Their study does not allow for efficiency gains from variable time-of-day pricing. However, their earlier study, "Optimal Peak-Load Pricing for Local Telephone Calls," R-3404-RC (Santa Monica, Calif.: Rand Corporation, March 1987), concludes that even time-of-day pricing may not lend to welfare improvements over flat-rate pricing if there are substantial variations in demand within periods, additional capacity costs are low, and measurement costs are substantial.

59. Mitchell, "Incremental Capital Costs"; and Rohlfs, "Marginal Costs."

opportunities, particularly in the urban areas, the second-best rationale appears dubious. As was shown in table 5-3, discrimination against business customers has in fact lessened in the 1980s in the largest exchanges in urbanized states, but not elsewhere. Businesses still pay more than double the residential rate in most jurisdictions. Given the relatively low elasticity of demand for access and the limited business-residential repricing, the welfare gains from this limited pricing reform are surely less than $100 million a year, even assuming no externalities in consumption.

REDUCTION OF THE URBAN-RURAL DIFFERENTIALS. The direct subsidization of rural subscribers through lower monthly access rates than are available in urban areas has also declined in the 1980s (table 5-5). Once again, however, the changes have been modest particularly in light of the very large differences in the fixed costs of service between rural and urban areas. In general, residential subscribers in the largest exchanges continue to pay more than 20 percent more for service than those connected to the smallest exchanges even though the capital costs in the latter exchanges are perhaps twice as high as those for the larger urban exchanges.[60] In December 1980, they had paid approximately more than 40 percent more for service. Given the low price-elasticity of demand for access and the small reduction in the rural-urban distortion, the improvement in social welfare from this repricing is once again quite small.

INCREASED SERVICES AND EQUIPMENT DIVERSITY. Surely, one of the most important contributions of competition in telecommunications has been the explosion of services and equipment offerings. Even without liberalization, AT&T and the independent companies might have developed or allowed others to develop the variety of telephone sets, data- communications terminals, answering machines, cellular equipment, and facsimile machines now available, but it may have been at a slower pace. Similarly, the wide variety of packet-switched and other value- added services, voice mail, long-distance packages, and other new services may have developed under monopoly, but the spur of competition may well have accelerated these developments.

A study of the rate of diffusion of new services and equipment in countries with liberalized or monopolized services and equipment

60. Mitchell, "Incremental Capital Costs" (estimate of differences in fixed costs for small and large exchanges).

markets should be undertaken in the next few years. Currently, such a study is beyond the scope of this book.

Summary

Undoubtedly, deregulation, divestiture, and the liberalization of entry have placed enormous pressure on regulators. The Federal Communications Commission has launched an effort to reduce the burden of local subscriber connection costs on interstate long-distance calls through the imposition of subscriber-line charges (SLCs) on final subscribers. These SLCs and the regulatory lag in catching up with the 1970s inflation have raised local monthly telephone rates greatly, particularly in the more rural states. But the evidence does not support the theory that overall telephone rates have been increased by regulatory policies or by the AT&T divestiture.

The rise in local rates and the consequent decline in interstate long-distance rates have had a mildly regressive impact on income distribution. The rising local access rates have led to the imposition of more measured local service options and Lifeline services to offset their impact on various consumers. There is no evidence that low- income consumers have discontinued their telephone service in significant numbers, but the share of the poorest households with telephone service would probably be about 2 percentage points higher if local rates had remained at 1980 real levels.

Early concern about declining quality of service and increasing riskiness of telephone company equities has proved unwarranted. Quality apparently has not declined. Nor has the systematic riskiness of the equities of the telephone service companies risen since the 1960s. In short, divestiture has caused surprisingly few disruptions.

The results in this chapter confirm the earlier observations about the shift of telephone service from regulated carriers to private business networks. By 1988 the regulated carriers' output was 25 percent below what would have been predicted from the pre-1981 relationship between output, relative prices, and GNP. Thus, liberalization seems to be offering businesses and consumers choices that were not available before the 1980s.

The efficiency of telecommunications service has improved much more rapidly than before liberalization. By 1988 this improvement alone was worth $3.5 billion a year, and a big share of it may be attributed to liberalization of entry. Offsetting these gains was a one-time adjustment cost to divestiture of $3.7 billion. However, the improvements in efficiency continue to grow while the costs of divestiture are now history.

Besides the gains in productive efficiency, liberalization has spawned great improvements in regulated rates. Most important, the repricing of toll and access services through the FCC's subscriber-line charges contributed $0.7 billion to $1.4 billion to economic welfare by 1988. The full implementation of this system could generate as much as $3.1 billion in annual welfare gains.

The other distortions in telecommunications rates, such as the overpricing of business service and the underpricing of rural residential service, have not yet been attacked seriously by regulators. The threat of bypass is having some effect on business rates in more urban areas and the direct subsidy to rural users is being reduced slowly. However, these changes are so gradual as to be of only minor significance at this juncture.

Finally, it is impossible in this book to quantify the benefits from new equipment or services. These benefits could be substantial given the rapid growth of new products available to residential and business subscribers.

Chapter Six

A Concluding Assessment

THERE IS NO DOUBT that the telephone industry has undergone a radical transformation in the past two decades. Even the smallest residential customers can now buy their own equipment, order service from dozens of vendors, and choose from among a dazzling array of special options. This diversity has its cost. No longer may the customer simply contact his or her Bell operating company for end-to- end service. Indeed, Bell operating companies are not even permitted to offer many services or to produce telephone equipment.

For the large business customer, the choices are even wider. Interstate long-distance service may be procured from AT&T or from several competitive carriers (OCCs). Equipment may be purchased from scores of different companies. A franchised monopoly probably still provides local service, but much of what was once local service may be obtained through private branch exchanges, local area networks on the customer's own premises, smart buildings operated by the landlord, and even large private (unregulated) fiber-optics networks. The user may even build a customized private network and choose from among scores of potential locations to interconnect with the public network.

In this new environment, many worry that network efficiencies have been sacrificed to the appetite for competition created by public policy. Others fear that the entire telecommunications system will begin to fragment because of a lack of common standards for interfacing competitive equipment to various public or private networks. And still others worry that the research and development that Bell Laboratories performed so ably for decades will no longer be underwritten by firms greedy for short-term profits.

Clearly, competition and divestiture have affected telephone rates, equipment prices, productivity, and quality of service. Competition has probably spurred technical efficiency in the telecommunications sector, telephone service has not collapsed, and divestiture may have caused at most a modest short-term increase in overall rates. The telephone rate structure, however, has changed markedly. Higher local access rates, especially in rural areas, are more efficient, but may impose some burden on the lowest-income subscribers. More important, the network is fragmenting as private networks and equipment reduce business usage of the public network. And the growth of regulated telephone common carriers has slowed dramatically in the 1980s.

Several serious, potential problems lie ahead in the more competitive world of telephony, stemming from the greater fragmentation of the network and the unfortunate combination of regulation and relaxed entry conditions in the industry. At this juncture, however, one can only speculate on the extent of these problems and any possible solutions.

The Cost of Fragmentation: Incompatibility

The telephone network began to fragment in the mid-1970s when the Federal Communications Commission (FCC) admitted specialized interstate carriers and required local telcos to allow the connection of foreign equipment to their lines. Thus several years before the AT&T divestiture, there were already pressures on network standardization.

When AT&T dominated the nation's telephone system, it essentially dictated the technical standards for the entire network. Today, however, AT&T no longer enjoys this dominant position. The seven regional Bell holding companies, several large independent telephone companies, the large interexchange carriers, and the principal equipment manufacturers may choose their own standards for any of a multitude of services. The engineer's nightmare is that equipment designed for one operating company may be incompatible with equipment designed for use on other systems. One solution to this problem is referred to generically as open architecture, that is, allowing downstream users to connect their equipment to the carrier's system at well-specified locations, perhaps through the use of adapters or translators.

This open architecture solution seems to have worked very well for standard customer premises equipment (CPE). No one complains that he cannot connect a telephone handset, modem, or even private branch exchange to an operating company's lines today. The transition from monopoly to competition in CPE has worked very well indeed as shown by the market-share statistics in chapter 4.

The open architecture solution is more controversial when it applies to sophisticated service offerings that require the local franchised monopoly's switched services to originate and terminate calls. This is particularly true for those services, such as electronic mail, call forwarding, and so on, that can be offered as a joint product with other telco services through adaptations of the sophisticated electronic switches currently being produced. A telephone operating company could conceivably configure service to make competition in these downstream services very difficult.

Attempts by the FCC to impose open network architecture on franchised local telephone monopolies has been fraught with controversy. The FCC has not told these telephone companies how to configure their systems to make them truly open. Perhaps such a prescription would be impossible. Yet, when the operating companies describe their proposed open network structure, they are met with criticism from service vendors who worry that the network is being designed to disadvantage actual or potential competitors.

Unfortunately, even if all parties could agree on the ideal open network architecture, the problem remains of determining how the franchised telephone monopoly should charge for the various types of services or connections demanded by downstream vendors. If the Bell operating companies are increasingly allowed to reenter downstream "information" services, this option will become a very contentious issue, reminiscent of the debate over access tariffs during the early days of MCI. The FCC and state regulators have never been able to disentangle joint and common costs of telephone service to anyone's satisfaction; therefore, regulation of relative rates is likely to be very inefficient.

It must be admitted that competition may cause a loss of standardization to occur. Any attempt, however, to impose standards from outside the industry runs several risks.[1] The regulator may be

1. See Stanley M. Besen and Garth Saloner "The Economics of Telecommunications Standards," in Robert W. Crandall and Kenneth Flamm, eds., *Changing the Rules: Technological Change, International Competition, and Regulation in Communications* (Brookings, 1989), pp. 177–220.

captured by opportunistic industry participants who argue for a set of standards that disadvantage their competitors. Or, alternatively, a standard may be imposed too soon, locking the industry into an inefficient technology. But even if a loss of economic welfare occurs because of uncertainty over network standards or even outright incompatibility, this potential loss must be weighed against the substantial improvements in productivity in the service and equipment sectors of the industry as well as against the remarkable increase in diversity of telephone services that has occurred since customers were freed from the grip of the telephone monopoly.

Research and Development

Under the old AT&T license agreement, the Bell operating companies were required to contribute a license fee, based on their revenues, to the corporate parent. Part of this license fee was used to support Bell Laboratories. With divestiture, this arrangement was obviously abandoned. Bell Laboratories may no longer rely on a tax on operating companies to support its activities.

At divestiture, a new research organization, Bell Communications Research (Bellcore), which drew heavily on the personnel of Bell Laboratories, was established for the divested regional holding companies. Thus, two principal national telecommunications research organizations now exist—Bellcore and Bell Laboratories. Neither, however, is supported by an automatic levy on telephone ratepayers. Bellcore's owners participate voluntarily in supporting its activities. Nor does either organization now have the luxury of undertaking research that is not relevant to the telecommunications business of its parents. The research activities of both organizations are much more focused on commercially practical ideas, rather than on long-term scientific research.[2]

Some observers had feared that divestiture would lead to a decline in research on telecommunications in the United States, but the results have not borne out this fear. In fact, real expenditures on telecommunications research have increased dramatically since divestiture. Data collected by the National Science Foundation for the years between 1972 and 1987 (table 6-1) show that although

2. See A. Michael Noll, "Bell System R&D Activities: The Impact of Divestiture," *Telecommunications Policy*, vol. 11 (June 1987), pp. 161–78.

Table 6-1. Expenditures on Research and Development in Communications Equipment (SIC 36), 1972–87
Millions of current dollars

Year	*Current dollars*	*1982 dollars*[a]
1972	2,583	5,555
1973	2,613	5,279
1974	2,424	4,489
1975	2,385	4,022
1976	2,511	3,979
1977	2,725	4,049
1978	2,999	4,154
1979	3,635	4,625
1980	4,024	4,695
1981	4,758	5,062
1982	5,758	5,758
1983	7,113	6,846
1984	8,354	7,757
1985	8,862	7,991
1986	9,023	7,929
1987	9,538	8,124

Source: Data obtained from the National Science Foundation.
a. Deflated by implicit GNP deflator.

research expenditures in the communications equipment industry declined in real terms in the 1970s, they rebounded sharply in the 1980s. Since 1977–78, the real amount of resources devoted to research in this industry has nearly doubled. Much of this increase, however, may be due to government funding of defense-related communications.

Nevertheless, "Bell System" research has increased rapidly since divestiture. AT&T outlays on R&D have increased steadily to about $2.6 billion in 1988 and Bellcore expenditures currently approach $1 billion (table 6-2). Thus, in nominal dollars, the fragmented Bell System now spends about twice as much on research and development as it spent in 1982 even though its revenues are only 60 percent above 1982 levels.

Even if total spending on research and development in telecommunications has not declined as a result of the divestiture, some analysts express concern that the vertical fragmentation has sacrificed potential economies of scope in the applications of new technology and reduced the ability of the divested regional holding

Table 6-2. Bell System Expenditures on Research and Development, 1972–88

Millions of current dollars

Year	*AT&T revenues*	*AT&T research and development outlays*	*Regional holding companies' revenues*	*Regional holding companies' payments to Bellcore*	*Research and development as a share of Bell System revenues*[a]
1972	20,904	657	...	...	0.031
1973	23,527	542	...	...	0.023
1974	26,174	584	...	...	0.022
1975	28,957	620	...	...	0.021
1976	32,816	641	...	...	0.020
1977	36,495	718	...	...	0.020
1978	40,993	838	...	...	0.020
1979	45,408	978	...	...	0.022
1980	50,658	1,190	...	...	0.023
1981	58,066	1,517	...	...	0.026
1982	65,866	1,790	...	...	0.027
1983	70,319	2,221	...	...	0.032
1984	33,188	2,368	57,937	728	0.034
1985	34,417	2,228[b]	63,319	783	0.031
1986	34,213	2,278[b]	67,257	816	0.030
1987	33,768	2,453[b]	69,766	842	0.032
1988	35,210	2,572[b]	74,288	905	0.032

Source: Annual reports.

a. Excludes regional holding company (RHC) non-Bellcore research and development. Note that total revenues of AT&T and RHCs include "double counting" of access charges paid by AT&T to the RHCs. Therefore, the 1984–88 share for research and development is biased downward when compared with 1972–83.

b. Includes all software development expenditures, which are now capitalized and amortized.

companies to appropriate the benefits of research.[3] These critics contend that foreign postal, telegraph, and telephone authorities do not suffer from these handicaps and are thus more likely to fund new communications technologies whose applications span the range of local, interexchange, and information services. It may be argued, however, that the guaranteed markets of a protected vertically integrated monopolist are less likely to induce aggressive technological development than are competitive, vertically fragmented markets.

At least one other line of argument is used against the vertical

3. Robert G. Harris, The Implications of Divestiture and Regulatory Policies for Research, Development, and Innovation in the U.S. Telecommunications Industry, Business and Public Policy Working Paper (University of California, Berkeley, December 1987).

fragmentation of the U.S. telephone industry in a world of rapid technological change. The regional holding companies are now forbidden to engage in manufacturing or in developing new communications services in the United States (other than a few voice storage or information gateway services). In the future, some of these restrictions may be removed, however, allowing them to begin to compete with AT&T in important U.S. markets. Because the regional companies are essentially unrestricted in foreign telecommunications ventures, they are impelled into joint ventures with foreign service companies and equipment suppliers. Moreover, they have an incentive to avoid AT&T as a supplier because of AT&T's status as a potential future competitor. The modified final judgment decree (MFJ) is thus inducing the divested Bell operating companies to favor foreign equipment suppliers, thereby adding to the U.S. balance-of-payments problem in a world in which foreign postal, telegraph, and telephone authorities continue their parochial buying habits.[4]

Finally, AT&T and the regional Bell holding companies now scrutinize research projects from the narrow perspective of self-interest. This result has created concern over the loss of Bell Laboratories as a great national research institution, but the solution to this problem is to find a source for funding important societal research from something other than a regressive tax on local telephone operating companies.

It is simply too early to muster the evidence on the effects of divestiture on technical change in the U.S. telecommunications sector. Lead times in the development of new technology are very long. Replacement cycles in capital spending on the network are also protracted. Measures such as the percentage of digital lines are not accurate indicators of the rate of technological progress. As discussed earlier, technical progress in telecommunications equipment has historically lagged badly behind that of the computer industry. Moreover, productivity growth in the regulated sector has not declined since divestiture. If competition in equipment markets is more important to innovation than the increase in appropriability of benefits from full or partial vertical integration, in a few years an acceleration in technical progress in the industry should occur.

4. But, as chap. 4 has shown, most of the decline in the U.S. balance of trade on telecommunications equipment lies in customer premises equipment, not transmission and switching equipment.

However, if economies of scope and appropriability problems are more important, the MFJ may well turn out to have impeded technological progress. Remember, however, that even though Bell Laboratories was viewed with great respect before divestiture, Western Electric was not seen in the same way. The large-scale retrenchment at AT&T and the big reductions in employment that followed the MFJ are evidence that Western Electric was not at the competitive frontier in all of its products.

Private Networks and Bypass of the Local Network

Since divestiture, policymakers and local operating companies alike have been seriously concerned about the threat of bypass of local telephone company circuits. Given that a large share of local company costs are fixed, any reduction in revenues for the use of their circuits requires that regulatory "revenue requirements" be made up from other services, such as local access-exchange services. If long- distance carriers must pay relatively high access charges for the use of local-company originations or terminations, they will search for less expensive means of reaching their customers. As long as the additional cost of originating or terminating calls through the local-exchange carriers (LECs) is less than the average cost of building new bypass facilities, economic efficiency requires that the calls pass through LEC facilities. If they do not, uneconomic "facilities" bypass occurs.

In the first three years after divestiture, the regional Bell operating companies funded numerous studies of the bypass threat. These studies typically showed that many large users were able to bypass the ordinary switched-access services offered by the local-exchange carriers, but that very little of this bypass was facilities bypass. Rather, these large customers simply negotiated for a lower rate, generally on a dedicated private LEC line, a phenomenon referred to as "service bypass." In 1988 facilities bypass was 8.7 percent of carrier access revenues.[5]

The Huber Report, submitted by the Department of Justice to Judge Harold Greene at the first triennial reconsideration of the line-of-business restrictions on the Bell operating companies, found

5. See chap. 3.

that although non-LECs had a lot of capacity for bypass, only 0.5 billion of 340.5 billion interexchange carrier minutes of use were originated or terminated on non-LEC lines.[6] This finding convinced Judge Greene that the LEC bottleneck remains and that the restrictions on the Bell operating companies should remain.

Unfortunately, much of the discussion of bypass ignores the obvious LEC bypass that is occurring—the use of private facilities by larger customers. It is impossible to measure how much interexchange traffic or termination of this traffic through LECs has been lost to private carriage. Nor is it possible to conclude that such private communications are inefficient in the sense that they are more costly than the incremental costs of providing the same service by way of the switched public network. The definitive analysis of this avoidance of the public switched network would require data from private users that are not generally available to the public.

The growth of private carriage in this regulated industry shows that either regulated rates are set inefficiently or that the economies of scale and scope justifying regulation are much weaker than recent econometric studies suggest. Conceivably, telecommunications meets the requirements for an "unsustainable" natural monopoly, but this theory has not been demonstrated empirically. Economies of scope exist in the delivery of local-access services and other services, but as Leonard Waverman has shown, such economies do not require that the same entity provide local access and these other services.[7]

In previous decades, the regulation of trucking and railroads so distorted rates as to drive a very large share of traffic into private truck carriage. Since the virtual deregulation of trucking in 1980, shippers have reduced their reliance on private carriage.[8] Presumably the reduction occurred because common carrier rates have fallen enormously, thereby reducing the profitability of a shipper owning a fleet of trucks. Were telecommunications rates to be related more directly to costs, private telecommunications carriage would probably also decline.

6. Peter W. Huber, *The Geodesic Network: 1987* Report on Competition in the Telephone Industry (Department of Justice, 1987).

7. Leonard Waverman, "U.S. Interexchange Competition," in Crandall and Flamm, *Changing the Rules*, p. 84, table 8.

8. Kenneth D. Boyer, "The Costs of Price Regulation: Lessons from Railroad Deregulation," *Rand Journal of Economics*, vol. 18 (Autumn 1987), pp. 408–16.

The Uneasy Coexistence of Competition and Regulation

Many people believe that telephone services have been deregulated, but in fact precious little deregulation has taken place. Local rates remain regulated. State commissions still regulate intrastate toll rates. The FCC regulates dominant carriers, such as AT&T. Only the provision of customer premises equipment has been truly deregulated and removed from the rate base.

The principal change in telephone regulation over the past two decades has been the liberalization of entry. There are now competitive suppliers of virtually every type of service and equipment. There are even some competitors for local access services, such as cellular services and private fiber-optics networks in large cities. Unfortunately, even when service competition is vigorous, regulators limit dominant carriers' pricing discretion and attempt to judge the reasonableness of costs and rates.

Two big problems plague the combination of regulation and liberal entry policies. First, regulators are likely to be forced into the role of cartel managers, particularly if one or more carriers find themselves unable to compete.[9] New entrants or even incumbents may use the regulatory process to prevent their rivals from cutting rates, claiming that such rate reductions are "predatory." This accusation occurs routinely in the interstate interexchange market whenever AT&T announces a reduction in rates or a rate concession to a large customer.[10] The recent decision by the FCC to allow AT&T to reduce rates within certain boundaries (as long as rates are above average variable costs) may mitigate but not eliminate the potential damages from this form of regulated competition.

Second, regulators may tighten barriers to entry in other markets to preserve some politically motivated cross subsidy. Obviously, entry is most likely to occur in the markets in which regulators have set rates very high relative to costs. Regulators, particularly at the state regulatory commissions, view such entry as a threat to the

9. See John Haring, "The FCC, the OCCs, and the Exploitation of Affection," OPP Working Paper 17 (Washington: Federal Communications Commission, June 1985).

10. For an example of this type of challenge, see any of the various disputes involving AT&T's tariff 15.

rate structure they have crafted in response to political pressures. They respond by protecting those markets remaining under their control from the additional entry that would contribute to a further unraveling of cross subsidies.[11]

For twenty years, the FCC has been pressing liberalization of entry in telephony, and the states have been resisting it. Whenever federal regulators are able to preempt state restrictions on competition, they have usually done so.[12] Local and intrastate toll services, however, remain under state control. Or do they? With current technology, how does a state public utility commission know when an apparently interstate communications firm is providing intrastate service? This firm may switch a message outside the state to connect two residents in the same state. Certainly, large business networks are already doing this switching. When will smaller customers be afforded similar "intrastate" service by competitors who simply evade state entry restrictions?

To their credit, the FCC and some state regulators realize that continued regulation of rates in markets in which entry is permitted is likely to create serious inefficiencies. As a result, they are searching for regulatory rules that might allow greater pricing freedom and provide more incentive for productive efficiency. The most popular candidate for replacing rate-of-return regulation is a price cap for a regulated firm's rates, which would limit the regulated firm to an average annual rate increase equal to the rate of inflation less a productivity improvement factor. This mechanism is currently in use for British Telecom and has been adopted by the FCC for interstate services in the United States.[13]

The price cap is certainly no panacea, but it merits consideration. There are problems in selecting the appropriate inflation rate and productivity improvement factor. Also unresolved is whether the

11. In the case of telephone regulation, these subsidies are largely from dense interexchange routes to the rural subscriber. See chap. 5.

12. The limits of this policy were reached in the case of carriers' depreciation policies. The Federal Communications Commission attempted to present state control over common carrier depreciation charges for intrastate services, but was rebuffed by the courts. See FCC CC Docket 79–105.

13. Great Britain, Office of Telecommunications, *The Regulation of British Telecom's Prices: A Consultative Document*, issued by the Director General of Telecommunications (London, January 1988). See "Policy and Rules concerning Rates for Dominant Carriers," FCC CC Docket 87–313. It has also been recently adopted by California for intrastate telephone services.

utility should have unlimited freedom to vary any single rate as long as a weighted average of its rates satisfies the cap. And if there is such freedom over individual rates, how are the weights to be established?[14]

Clearly, the design of an efficient regulatory mechanism eludes telephone regulators. But so does full deregulation of most services that have been opened to competitive entry. At this juncture, analysis of the deregulatory option is urgently needed, for it is not obvious that continued regulation of interstate voice and data services does not reduce welfare by more than the potential monopoly pricing of a deregulated dominant carrier. The political resistance to deregulation obviously comes from rural subscribers in particular and residential subscribers in general. Competition is likely to be focused most intensively on business services on the most dense routes. These services will probably continue to subsidize rural residential service, though to a lesser extent than in the mid-1970s.

The Role of the Divested Bell Operating Companies

Under the MFJ, the divested Bell operating companies were to be restricted to monopoly services, with two important exceptions: Yellow Pages and the sale of terminal equipment. The Bell operating companies were barred from offering interexchange service and information services as well as from telephone equipment manufacture. Recent decisions by the district court overseeing the MFJ, however, have begun to admit the Bell operating companies into some information services markets.

As long as the Bell companies or other local-exchange companies have a bottleneck on local connections to all but the largest customers, they may use this bottleneck to frustrate competition if they are permitted to offer any services other than local access-exchange services.[15] Their incentive to behave in an anticompetitive fashion derives largely from their regulated status. In the absence of regulation, there would be much less reason to deny even a

14. For a recent discussion of price caps, see Timothy J. Brennan, "Regulating by 'Capping' Prices," Discussion Paper EAG 88-11 (Department of Justice, September 1988).

15. See Robert W. Crandall, "The Role of the U.S. Local Operating Companies" in Crandall and Flamm, eds., *Changing the Rules*, pp. 114–47.

monopolist the right to integrate forward or backward. With regulation, however, the local carrier may attempt to "fool the regulators" by shifting costs from the competitive market to the regulated market, thereby raising regulated rates and subsidizing its thrust into a competitive market, such as long-distance services.

Similarly, a regulated company that buys equipment from an unregulated subsidiary may either hide excessive returns in the upstream equipment supplier, be relatively indifferent to upstream inefficiencies in equipment manufacture, or both. The early results of divestiture suggest that the new vertically fragmented structure of the telephone industry has indeed created serious competitive pressures in equipment markets.

Policymakers are currently on the horns of an uncomfortable dilemma. On the one hand, by prohibiting the seven divested Bell regional holding companies from engaging in communications equipment manufacture, long-distance service, or information services, they may be forcing the public to forgo large benefits from either the potential economies of scope or simply greater competition. On the other hand, policymakers fear that the entire pre-1982 AT&T problem may reappear if the regional holding companies are allowed to reenter these forbidden markets.

There are three obvious potential solutions to this dilemma: forbid all participation by regional holding companies in the three forbidden markets until local-exchange service is truly competitive and deregulated; allow them to enter, but only with regulatory safeguards; or abandon rate-of-return regulation for local-exchange companies.

The first option—barring participation by Bell operating companies in other markets until local access is workably competitive—depends on an aggressive procompetitive stance in local service by state regulators and the FCC. Unfortunately, most state regulators still resist competition at the local exchange because such competition would obviously require cost-based rates.[16] Many state regulators are wed to a system of cross subsidies from toll to local service and from urban to rural customers. Competition and the repricing initiatives undertaken by the FCC are slowly reducing these subsidies, but much of the subsidies remains.

Nor has the FCC been an unwavering devotee of competition.

16. This is beginning to weaken. See chap. 3.

In the 1970s it allocated spectrum for only two cellular telephone systems in each market—one had to be reserved for the local franchised monopoly wireline carrier (LEC). Moreover, the FCC took seven years to begin licensing this spectrum after its initial allocation decision. Had cellular telephone been allocated only to competitive carriers and licensed much more expeditiously, the potential competition for the LECs might be far more vigorous today. Unfortunately, there were only 2 million cellular subscribers by the end of 1988—a paltry total when compared with the 125 million access lines provided by the LECs.[17]

The second option, the imposition of regulatory safeguards, has been pursued by the FCC for more than a decade. Among the choices considered and even adopted at various times have been separate subsidiaries and accounting rules. Under *Computer II*, the FCC initially required dominant carriers, such as AT&T, to operate separate subsidiaries for competitive businesses. This requirement became less compelling with the AT&T divestiture, and the FCC subsequently dropped it in favor of accounting safeguards in its 1986 *Computer III* decision.[18] Neither alternative is a satisfying approach to the problem. It is unrealistic to expect separate subsidiaries of the same firm to act independently of one another, and it is equally unrealistic to expect that regulatory accountants can divine rules that allocate joint and fixed costs reasonably, especially in the face of rapid technological change. For these reasons, Judge Greene has been reluctant to allow the Bell operating companies to reenter equipment manufacture or downstream services with only the FCC's accounting to protect against abuses.

The third option, the elimination of rate-of-return regulation, is most attractive to traditional economists familiar with the literature on the distortions caused by rate-of-return regulation. The most popular current alternative is the use of rate caps that limit the rise in rates to the rate of inflation less a productivity growth factor.[19]

17. See chap. 3, note 12.

18. In June 1990, the U.S. Court of Appeals for the Ninth Circuit overturned the FCC's Computer III decision, forcing the FCC to reconsider the issue later this year.

19. See, for example, Ronald R. Braeutigam and J.C. Panzar, "Diversification Incentives under 'Price-Based' and 'Cost-Based' Regulation," *Rand Journal of Economics*, vol. 20 (Autumn 1989), pp. 373–91; and Tracy R. Lewis and David E. M. Sappington, "Regulatory Options and Price Cap Regulation," *Rand Journal of Economics*, vol. 20 (Autumn 1989), pp. 405–16.

Rate caps are superior to rate-of-return regulation in providing incentives for efficiency as long as the productivity improvement factor is not readjusted periodically in response to actual carrier performance. But without such an adjustment, rate caps may be set too high or too low by regulators who can only guess at future improvements in technology.

At this juncture, rate caps appear most promising as a transitional step to full deregulation of interstate long-distance services.[20] The MFJ was structured on the basis of a theory that all telecommunications markets except local access-exchange could and should be competitive. Interexchange services have indeed become much more competitive since 1982. As equal access has spread to 90 percent of all operating company access lines, it is now possible to consider eliminating interstate long-distance regulation altogether. Rate caps, now in place for AT&T and for federal regulation of the local-exchange carriers, are a promising step in the direction of total deregulation of interstate long-distance services.

The Bell operating companies will probably not be allowed to enter inter-LATA (local access and transport area) interexchange markets or central-office equipment manufacture soon. Their equities have outperformed the general stock market since divestiture, making it difficult for them to obtain a sympathetic audience for their complaints of being confined to a declining industry. Consumer and elderly groups fear that reentry of the Bell companies into other telecommunications services will lead to further upward pressure on local rates. And Congress has been unwilling to consider a transfer of the authority from the district court administering the decree to the FCC. Thus, the Bell operating companies will undoubtedly find that they can only free themselves gradually from some of the line-of- business restrictions in the decree.

The Future of the Bell Operating Companies

While the MFJ line-of-business restrictions on the Bell companies seem to be politically invulnerable for the present, pressures will probably mount for some loosening of these restraints in the not

20. See John R. Haring and Evan R. Kwerel, "Competition Policy in the Post-Equal Access Market," OPP Working Paper 22 (Washington: Federal Communications Commission, February 1987), for a discussion of such a transition.

too distant future. Because of these restraints, the seven regional Bell holding companies are pursuing growth opportunities in other areas in which they are likely to have less comparative advantage. Unless these far-flung ventures prove more successful than recent diversification attempts by firms in other slow-growing industries, the growth of the regional holding companies may soon decline noticeably, exacerbated by the recent slowdown in local rate increases.

The Bell companies are most likely to succeed in freeing themselves from the information-services restrictions in the MFJ. Indeed, they have already opened a crack in this market with Judge Greene's decisions to allow them to provide gateways for videotex and to offer voice-messaging services. Given the slow pace of development of new services, such as videotex, the argument that the line-of-business restrictions on information services are necessary to prevent monopoly abuses of the local-access bottleneck rings rather hollow. It is difficult to reduce output to less than zero.

Unfortunately, introduction of new information services may require the development of hardware or software that delivers these services. The line-of-business restrictions preclude "manufacturing" by Bell operating companies—an activity that has been defined to include the design of equipment for exclusive production by outside contractors.

The Bell companies are also pressing hard for the right to offer fiber-optic cable services to all subscribers. Predictably, the cable television industry opposes this proposal, but a more general opposition may develop if "fiber to the home" results in higher average local loop costs and, therefore, higher basic telephone rates.

The relaxation of the manufacturing prohibition may prove more difficult. First, even though the data on price performance in the equipment industries are still sparse, the divestiture of Western Electric from the Bell operating companies will probably prove to have been among the most beneficial aspects of the MFJ. Second, pressures are mounting for an end to favoritism by telecommunications authorities toward national telecommunications equipment suppliers in Europe and in Japan. If formal divestiture and procurement liberalization lead to a more rapid pace of technological progress in telecommunications equipment, it will be difficult to persuade U.S. policymakers to allow the Bell operating companies a step back toward the status quo ante.

It now seems that the equal-access provisions in the MFJ may prove as successful in opening interexchange competition as the 1970s interconnection regulations for customer premises equipment. If a competitive interexchange market develops, the case for admitting the Bell companies into inter-LATA services will be less compelling. After all, why risk a return to the problems of the late 1970s if the current market is workably competitive? It may be possible, however, to admit the Bell companies into the offering of inter-LATA services for larger business customers or at least to allow them to contract with interexchange carriers to offer such service as part of a comprehensive telecommunications package to larger customers.

If cellular systems or less expensive personal communications networks become a meaningful alternative for local access by even dispersed small residential or commercial subscribers, the case for the line-of-business restrictions diminishes and even disappears. One option would be to divest the Bell companies' existing cellular operations in return for a relaxation of the line-of-business restrictions. This might increase the rate of diffusion of cellular and increase competition in the market for network access.

Ironically, the regulatory distortions in rates may increase the probability of relaxation of the line-of-business restrictions. If these rate distortions invite network bypass at an increasing rate, the recent slow growth of the Bell companies' telephone services will be reduced even further. This deceleration, in turn, will place pressure on the court and Congress to find mechanisms for relieving the squeeze on the Bell companies' operating margins by some mechanism other than local rate increases. One solution could be an expansion in the services they are authorized to offer—information services, interexchange services, or fiber-optics broadband service.

Conclusion

Even though the United States has adjusted well to the breakup of AT&T, the liberalization of entry into telecommunications markets, and the confusing array of choices now available in telephone services and equipment, formidable problems remain. Foremost among these problems is the continuation of regulation in markets with increasing numbers of competitive vendors. Telecommunications regulators cannot separate the costs of various services and therefore cannot efficiently regulate the complex array of services and rates, nor can they settle disputes among rivals about the fairness of rates.

The dangers in continuing to regulate an industry in which entry is permitted are well known. Regulators find themselves beholden to the pressure of incumbent carriers, especially when these carriers are required by the regulators to offer various nonremunerative services. If regulators continue to defend subsidies to rural subscribers and to attempt to cover non-traffic-sensitive, subscriber-line costs from other services, they will be forced to keep protecting incumbent carriers.

Despite the formidable regulatory problems that remain—many of them caused by the market liberalization policies of the FCC and the divestiture of AT&T—it would be difficult to ignore the numerous benefits stemming from the increase in competition over the past decade or more. Telecommunications may be an industry of economies of scale and scope, but it is also an industry on the cusp of the electronics revolution. Competition, not regulated monopoly, is likely to be the best market structure for encouraging technical progress.

Market liberalization in the telephone industry has placed increasing pressures on regulators to eliminate the regulatory distortions in rates that developed over the past decades. Interstate long-distance rates have fallen dramatically, partly because of an FCC decision to place a larger share of the costs of local access where they belong—in the monthly flat rates paid by subscribers for this access. Intrastate rates have fallen less rapidly because state regulators are less attracted to arguments about efficiency and more politically attuned to the demand for residential and urban service subsidies.

In the 1980s the price of local access-exchange service rose in real terms after falling throughout most of the 1960s and 1970s. Much of this rise in rates occurred because of regulatory lag as regulators allowed telephone companies belatedly to catch up with the inflation of the 1970s. Nevertheless, telephone equities have not recovered to their 1960s' market-book ratios.

The rise in local rates during the early 1980s clearly affected the willingness of low-income households to subscribe to telephone service. At most, these rates reduced telephone penetration in the mid-1980s by 1 to 3 percentage points compared with what it would have been, but the overall share of households subscribing to telephone service rose to more than 93 percent throughout the decade.

The repricing of telephone service—rising local rates offset by sharply declining interstate long-distance rates—had a regressive effect on income distribution. Lower-income households typically consume less long-distance service and fewer other services that are telephone intensive, such as financial and travel services. My estimate of the effects of repricing through 1987 find that lower-income households pay between $15 and $16 more a year because of repricing, while households with more than $40,000 a year in income save about $15 a year.

The net welfare gains from repricing, however, are great. Through 1988 the annual welfare gains from repricing ranged from $664 million to $1.4 billion. Were all rates, interstate and intrastate, to be set in Ramsey quasi-optimal fashion, the net welfare gains could be as much as another $10.9 billion a year. Hence, the attempt to use telephone rates as mechanisms for redistributing income is costly, reducing social output by more than two dollars for every dollar transferred from upper-income to lower-income households. It is far better to target subsidies to very low- income households through Universal Service funds or other mechanisms than to distort rates for all telephone subscribers.

The benefits of technical change in this industry have been enormous. Total factor productivity (TFP) growth has accelerated almost continuously since 1971 after remaining constant in the 1960s. Although TFP growth seems to have dipped during the tumultuous period of AT&T divestiture in 1984–85, it is now accelerating rapidly once more despite the fragmentation of the network owing to long-distance and local competition and the development of private networks. Employment in the telephone sector has fallen by about 30 percent since the announcement of the AT&T divestiture, reflecting the increasing pressures for efficiency in a more competitive world. By 1988 telephone industry costs were $3.5 billion lower than they would have been if productivity growth had remained at its 1960s' levels.

Nor have divestiture and competition reduced the industry's commitment to research and development. By the late 1980s, the Bell operating companies and AT&T were spending a larger share of their revenues on research and development than before divestiture.

The restrictions on the regional Bell operating companies are probably denying society the fruits of some economies of scope in telecommunications. These restrictions should be lifted as soon as the threat of cross-subsidies from the regulated local bottleneck monopolies disappears. State and federal regulators should be pressing policies to break the local monopoly of the traditional telephone carriers. Obviously, this action cannot occur as long as state regulators distort local and intrastate toll rates in response to political pressures.

The most difficult task in measuring the effect of liberalization and divestiture is trying to determine the impacts on equipment manufacture. The price data are simply too poor to draw definitive conclusions, and technological change makes all such analyses difficult anyhow. The equipment being produced in 1990 is so different from that produced in the early 1980s or the 1970s that measuring the gains in welfare is problematic. Facsimile machines, equipment used to construct local-area networks, and cellular telephones were unavailable in the early 1970s. Moreover, determining how much progress would have occurred in the absence of liberalization or divestiture is difficult.

On balance, the efficiency gains from the opening up of the telephone industry have more than offset the possible losses that may be caused by the sacrifice of economies of scale and scope or the absence of fully compatible equipment and services. As the telephone network fragments further, regulators will be forced to abandon the distorted rates that seem partly responsible for this fragmentation. Then it will become possible to get a market test of the magnitude of scale-scope economies versus the benefits of competition and new or customized services that have resulted from liberalization and divestiture.

Index